The Wise Advocate

The WISE ADVOCATE

The Inner Voice of Strategic Leadership

ART KLEINER, JEFFREY SCHWARTZ,
and JOSIE THOMSON

Columbia University Press
Publishers Since 1893
New York   Chichester, West Sussex
cup.columbia.edu

Library of Congress Cataloging-in-Publication Data
Names: Kleiner, Art, author. | Schwartz, Jeffrey, 1951– author. |
Thomson, Josie, author.
Title: The wise advocate : the inner voice of strategic leadership /
Art Kleiner, Jeffrey Schwartz, and Josie Thomson.
Description: New York : Columbia University Press, [2019] |
Includes bibliographical references and index.
Identifiers: LCCN 2018026093 (print) | LCCN 2018038810 (e-book) |
ISBN 9780231545860 (e-book) | ISBN 9780231178044 (cloth : alk. paper)
Subjects: LCSH: Leadership—Psychological aspects. |
Management—Psychological aspects.
Classification: LCC HD57.7 (e-book) | LCC HD57.7 .K548 2019 (print) |
DDC 658.4/092—dc23
LC record available at https://lccn.loc.gov/2018026093

Columbia University Press books are printed on permanent
and durable acid-free paper.
Printed in the United States of America

Cover design: Noah Arlow

Contents

The Wise Advocate

CHAPTER ONE

The Choice of the Moment

HAVE YOU EVER had a difficult executive decision to make? This is the kind of decision where the best options aren't obvious, the consequences aren't crystal clear, and the outcome could affect hundreds of people or more. Each of these moments of choice is a potential turning point—it may move your organization, or you personally, across a threshold you haven't crossed before.

Maybe you're the leader of a company. You have a great innovation in mind, but it would mean taking on so much debt you'd practically be betting the company on this venture. Or you're asked to oversee a round of layoffs, and you must decide who stays and who goes. Or you're an up-and-coming entrepreneur. Someone offers you a lucrative deal, but you must change the way you operate, and you're not sure you and your colleagues are up to the task. Or you are asked to cross an ethical line: to pay a bribe, adjust some data to make the outcome look better, or hire a friend or relative when there are other, better candidates.

Some of these difficult choices may come up outside the workplace as well. In deciding which job to take, where to send your children to school, whether to change your home, how to handle a debilitating disease, what to do about a struggling marriage, how to think about a relative or friend who is under great stress, or in many other life situations, you may find yourself facing a dilemma you never planned for, deciding among uncertain alternatives. Sometimes it may feel as though the only expedient choice is morally questionable; at other times, you may be desperate and feel as though you have no choice at all. At still other times—at work or at home—a dramatic new opportunity may beckon, but you feel timid about it. You don't know exactly what will happen, and you have to take a leap of faith no matter what you do; even standing still and not taking any action is a sort of decision.

In the face of such uncertainty, how do you figure out the right thing to do? Once you figure it out, how do you set yourself up so that things turn out well after you make the choice? Most important, how do you develop the habit of making better choices, time and time again, even in difficult and uncertain circumstances?

No book can answer these questions completely, of course. The best choice is always specific to the situation. But all major decisions—whether a significant business move or a deliberate life change—have one thing in common. At the moment of making the choice, certain things happen in the human mind and brain that can, if you pay attention appropriately, help you make better decisions in the future. Moreover, if you manage this type of decision in the right way, time and time again, you can become one of those rare people known as strategic leaders—people who can help a group, an enterprise, or a country transcend its limits and move closer to fulfilling its most significant aspirations.

Neuroscientists and psychologists have studied the mind (the locus of mental activity) and the brain (the physical organ associated with that activity) in ways that illuminate these critical moments of choice. The results of their studies—along with insights from organizational research and observation—have illuminated the neural dynamics of people in leadership roles. Insights about neuroscience and leadership are emerging from many places, and they all point to the same relationship among decision-making habits, recurring mental activity, and this quality of strategic leadership—the ability to address problems that are too complex for an organization to manage easily. A strategic leader gains legitimacy by helping the organization transcend its limits and move toward more profound and powerful levels of capability.

This requires being able to inspire others, help groups consider new possibilities, and bring enterprises toward lofty, beneficial goals.

In this book, the relationship between the human mind and brain is central to strategic leadership. We define the mind as an ongoing pattern of mental activity associated with the self, and especially with the conscious choices and decisions made by any human being. The brain is the biological organ in the skull where neurons and other cells process and transmit sensations, feelings, thoughts, and emotions. The mind is active—the source of choices and decisions you make about how and where to focus your attention. The brain is affected by these choices and decisions, and the more capable you become at recognizing and managing your brain's reactions, the easier it becomes to focus your attention effectively. That leads to the simple concept at the heart of this book: *The focus of your attention determines what happens in the mind and the brain during these critical moments of choice, and determines what kind of a leader you can be.*

As we explain this concept, we will also explore some fundamental questions—questions that are still the subject of ongoing active neuroscience research—about the mutual influence and impact of the brain and mind on each other. But all this complexity comes down to the fact that in most instances, you are likely to focus your attention in one of only two basic ways. We call them the "Low Ground" and "High Ground," and they can lead you, respectively, toward transactional or strategic leadership.

The Nature of Strategic Leadership

The concept of strategic leadership has a rich, broad conceptual lineage that goes back to Homer and Plato. In recent years, it can be traced to General George C. Marshall and his successors at the U.S. Army War College;[1] to the seminal leadership writer James McGregor Burns, with his concept of transformative leadership;[2] to a group of writers positing that some leadership is grounded in innate character traits, most notably Bill George with authentic leadership and Ronald Heifetz with adaptive leadership;[3] and to many others, including Warren Bennis.[4] These theories all share two core themes that we think are significant for anyone who is trying to have an effect on the world around her or him.

First, the pattern of mental activity—or, more loosely, the way you think and behave—matters. This starts with the observation, obvious but profound, that different people placed in the same role—at the top of a government, an enterprise, a church, a group, or a country—can lead the organization down very different paths. Moreover, a successful leader of one organization will often (but not always) generate success at another. Why, then, do some people lead groups to thrive, while others take them to decline? The history of leadership theory is a search for answers. What they all have in common is the importance of the leaders' mental activity, manifest in what they say, what they do, and most subtle but most important, how and where they focus their attention.

Second, there seem to be two broad patterns of successful leadership. The first is what Burns calls transactional leadership. He defines it as "exchanging gratifications"; for instance, he writes, the success of the Democratic party in the United States under Franklin D. Roosevelt in the 1930s depended in part on the exchange of loyal support and votes for benefits like Social Security.[5] The leader retains legitimacy so long as the wants and needs of the followers (or of other deal-makers) are gratified. A gifted transactional leader, like the heads of many corporations, is particularly skilled at solving the problems of the moment. Ron Heifetz and Marty Linsky call them "technical challenges";[6] they already have clear answers, and solving them is primarily a matter of executing them with expedience—being efficient, pleasing stakeholders, and making the boss and main constituents happy.

But few of us get to deal only with problems of the moment. For most leaders, in most organizations today, problems are more complicated. Heifetz and Linsky label them "adaptive challenges." They're also known as "wicked problems," a term coined by Wes Churchman, or as VUCA situations, an acronym that the U.S. military has used since the early 1990s to describe situations with volatility, uncertainty, complexity, and ambiguity.[7] Here, there is no obvious solution, and the leaders have to look freshly at the situation.

This is the domain of strategic leadership, which Burns called "transforming" or "transformative" leadership. It is also associated with other leadership concepts—among them, Bill George's authentic leadership and Ron Heifetz's adaptive leadership. Indeed, there is a broad literature about strategic leadership, full of observations and biographies of leaders who have broken through habitual patterns to achieve extraordinary results.[8]

Strategic leadership typically involves the type of group dynamics in which everyone understands that the problem is more difficult than the skills and capabilities of the organization, at least as they are currently organized, can manage easily. Something significant has to change. Some current ways of thinking have to be challenged. The strategic leader is influential within the group. He, she, or they (there may be more than one) is the person who believes, and can inspire others to believe, that the problem is worth solving; that it cannot be solved in conventional ways; that these individuals can solve it if they're willing to change the way they look at things; that a new way of looking at things is available to them; and that they can all make a difference together if they act on this belief.

One might think of these leaders as Wise Advocates; they are voices within the enterprise speaking and acting on behalf of longer-term, broader-perspective goals—the aspirations that led people to join that organization in the first place. Often strategic leaders have earned positions of authority; they may be CEOs or board members. But they don't have to be at the top of the organization; they might, for example, hold a key functional position such as chief strategy officer, head of finance, or director of human development. Moreover, even if they are chief executives, their positions are not the only source of their legitimacy. Strategic leaders gain influence and impact because of what they do and say consistently, on behalf of the direction of the whole. Their voice, their way of consistently articulating a Wise Advocate position, is what makes them effective and influential as leaders.

Strategic leaders can play the role of Wise Advocate for their organizations only because they have similar Wise Advocates in their minds—mental constructs that they have cultivated over time. To call on your Wise Advocate is to take a third-person perspective on your first-person experience—to see yourself, and the way you interact with the world, as a trusted, caring, and dispassionate observer would see it.

This inner voice of strategic leadership, which we will consider at length in this book, is a well-known mental phenomenon, associated with foresight, clarity, and virtue. Its existence is also consistent with what we know about neuroscience and the mental activity associated with maturity, executive function, contemplative self-awareness, and some forms of empathy. When practiced over time, it can become so ingrained in the mind that it is etched into the physical makeup of the brain. Leaders who actively consult their Wise Advocate are less daunted when they face a truly challenging situation. They have the habit of taking a broader perspective, of looking for the right solution, not just the people-pleasing or transactional solution. They may not be self-conscious about it; over time, it has become their normal way of making tough decisions and a natural aspect of their leadership presence.

Low Ground and High Ground

How do you develop this quality of leadership in your own mind? Let's go back to that moment of decision we described at the beginning of this chapter: the ambiguous choice, where every option has tradeoffs. You might focus on expedience: on getting what you want and giving others what they want, as rapidly and efficiently as possible. This leads you to a powerful pattern of mental activity that often makes it possible to solve problems—but generally in a transactional, direct way. It will tend to satisfy the urgent-seeming demands that your brain produces and the demands that others place upon you. In this book, we refer to this pattern as the Low Ground of the mind, and we associate it with particular circuits within the brain. Because the Low Ground feels natural, it often becomes habitual for leaders. That strengthens the brain circuits related to it and makes it harder to resist.

But you may have reason to resist. At that moment of decision, you might choose to move toward outcomes that transcend the needs of the moment, that fulfill what the situation truly calls for—the best long-term consequences, based on the values most important to you and your organization and on your perception of yourself and the other people involved. If you make that choice, then that will lead to a different pattern of mental activity, which we call the High Ground. It, too, is associated with particular brain circuits. The more you follow this path, the easier it is to transcend your limits and the limits of your organization and to shape your future accordingly.

The Low Ground is tactical and transactional; the High Ground is strategic and transformative. When you are active in the Low Ground (or, more precisely, when your mind and brain are triggered by Low Ground patterns of mental activity), it feels as though you are acting from within the maelstrom of current pressures. This is often accompanied by powerful emotions, such as anxiety, fear, desire, craving, exhilaration, and the need for relief. The High Ground feels more like a path that you have chosen deliberately, often because you recognize the limits of conventional thinking and action. You are prepared to endure some doubt and discomfort along the way because you believe it will feel (and be) better in the long run.

When you spend time in the High Ground of your mind, it generally involves calling on the Wise Advocate—the inner voice of strategic leadership. This may feel like summoning a literal voice within yourself. Alternatively, it may feel like stepping back and being thoughtful: looking at your own situation, and yourself in it, with all the commitment and information of an insider but the detachment of an outside observer. As we said earlier, this means taking a third-person perspective on first-person experience. You care about the situation deeply; you are committed to the results you want; you are prepared to act decisively. But you are also dispassionate. You aren't trying to make everyone happy or to gain your own emotional advantage. You don't necessarily want to make anyone *un*happy in the short run, and you certainly don't intend to shortchange yourself or put yourself down, but you are willing to consider whatever is needed if it is a requisite part of a longer-term, broader-based positive outcome.

It's important to remember that the High Ground and the Low Ground are both states of mind, not aspects of your identity. They represent what you think, feel and do, rather than who you are. People switch between them, and to other patterns of mental activity, many times over the course of a day. Moreover, you always have a choice about which pattern to invoke, and you can learn to strengthen your ability to choose the High Ground when you want to.

Because of the nature of the human mind-brain relationship, people often forget that they have a choice. Especially when you invoke or inhabit the Low Ground, what comes to your mind during the day may not seem to involve choice at all. It often takes a form we will call "deceptive messages"—ongoing ruminations about your hopes, fears, anxieties, desires, and reactions to the outside world. These deceptive messages, as we'll see, govern a lot of individual reactions and are influential in the culture of most organizations. Left unnoticed and unchecked, they tend to draw people back toward expedience and the Low Ground. (They are also sometimes called "cognitive distortions.")

To move to a High Ground pattern of mental activity, you often have to make a conscious decision to reframe those deceptive messages and think differently about them than you ordinarily do. You have to deliberately call upon the Wise Advocate, the inner voice of strategic leadership within yourself, to help you do so.

Because they link mental activity and brain circuits, both the High Ground and the Low Ground are habit forming. Like all habits, they are not static; they change over time. Occupying the Low Ground can be addictive, for both you and your organization; it can lead you to assume, without ever reflecting on or testing the assumption, that the only good solutions are expedient and if you're not making yourself (or other people) immediately happy, you're not doing the right thing. The High Ground is habit forming in another way; as you become more proficient there, your confidence grows and your perspective broadens, and that makes you more likely to want to return.

The Low Ground and High Ground are associated with different parts of the brain, both within an area called the medial prefrontal cortex: the dorsal (higher) parts with the High Ground and the ventral (lower) parts with the Low Ground. When people hold their heads upright, the dorsal area sits above the ventral area in the brain. This is one reason that the names High Ground and Low Ground seem apt to us.

Our distinctions are different from the "high road" and "low road" distinctions about mental activity articulated by psychologist and neuroscientist Joseph LeDoux.[9] LeDoux's work has to do with different ways of processing emotional reactions. As we'll see later in this book, most leaders are mature enough to operate primarily on LeDoux's "high road"; they have learned to react thoughtfully rather than indulging their immediate impulses and emotional responses. Or as Daniel Kahneman might put it, they are adept at "thinking slow" as well as "thinking fast."[10]

But even with that level of maturity and emotional intelligence, they might still only occupy the relatively Low Ground of transactional leadership. In the long run, to be a strategic leader, they need to not just travel the high road as articulated by LeDoux, but spend time in the High Ground of the mind as well. You could say they need to practice High Ground mental activity until it moves from Kahneman's System 2 ("thinking slow") into the more automatic System 1 ("thinking fast") and becomes, in effect, second nature.

Few strategic leaders can stay on the High Ground all the time, but they know how to get there when they need to. The practice that helps them do this is multifaceted, integrating advances in four fields: psychology (the study of the human mind and human behavior), neuroscience (the study of the anatomy and physiology of the brain), management (the study of the governance and direction of human social systems), and ethical values (the practice of life in accord with deliberate choices). If the leadership hypothesis at the heart of this book is true (and it is consistent with current knowledge in all four of those fields), then the quality and effectiveness of your leadership depends on your ability to move deliberately from the Low Ground to the High Ground.

Figure 1.1 lays out the moment of choice we've just described in schematic form. Whichever way your decision goes, it affects four different types of activity: the thoughts accessed by your mind (deceptive brain messages or the Wise Advocate construct); broader patterns of mental activity (Low Ground or High Ground); the brain functions associated with those patterns (again, Low Ground or High Ground); and the type of leadership that your choice engenders (transactional or strategic).

The High Ground is also a transitional path to another pattern of mental activity that we call the Higher Ground. In this state of mind, which is not as easy to reach at first as the High Ground is, you are more completely conscious of your own attention, thinking, and awareness—and of your organization's subtle cultural dynamics. The Higher Ground enables a deeper, more transformational form of leadership, with the potential to foster change at a still larger scope and scale. We'll have more to say about the Higher Ground later in the book, as an explanation of the ways of thinking to which the High Ground can lead you.

Of course, the High Ground, Low Ground, and Higher Ground are metaphors, intended to make a complex subject more accessible. But they refer to real phenomena, in both the brain and the mind. Metaphors are an effective way to bring to life the abstract dynamics of mental activity. These particular metaphors are, to the best of our knowledge, consistent with existing knowledge of cognitive science and management research, and we have taken pains to keep them consistent even when it means making the narrative more complex.

For example, when we first conceived of this book, we limited our framework to the Low Ground and the High Ground. But as we began to discuss the concepts, test them against real-world examples, and consider responses from early readers of the manuscript, it became

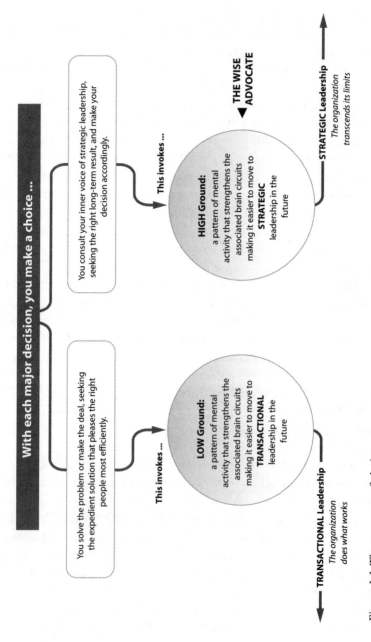

With each major decision, you make a choice ...

You solve the problem or make the deal, seeking the expedient solution that pleases the right people most efficiently.

This invokes ...

LOW Ground: a pattern of mental activity that strengthens the associated brain circuits making it easier to move to **TRANSACTIONAL** leadership in the future

— **TRANSACTIONAL Leadership** — *The organization does what works*

You consult your inner voice of strategic leadership, seeking the right long-term result, and make your decision accordingly.

This invokes ...

HIGH Ground: a pattern of mental activity that strengthens the associated brain circuits making it easier to move to **STRATEGIC** leadership in the future

▶ **THE WISE ADVOCATE**

STRATEGIC Leadership — *The organization transcends its limits*

Figure 1.1 The moment of choice.

clear that these two patterns of activity were not adequate to describe the dynamics we were trying to represent. This compelled us to add the Higher Ground. (As we'll see, there is also a Lower Ground beyond the Low Ground, triggered by sensation and instinctive emotion, equivalent to LeDoux's low road. But again, most business decision makers have learned to move past the Lower Ground, at least in their professional lives.)

Because the High Ground is a gateway to the Higher Ground, while the Low Ground can be addictive, it would be easy to say that the High Ground is good for you and the Low Ground is bad. But we do not condemn the Low Ground per se. Expedience can often be useful. The Low Ground is the path you might take when you feel the need to ease tension and pain, or to make a fair deal. It leads you to put your attention on creating immediate value: solving problems, getting results fast, making people feel better, and giving everyone (including yourself, sometimes especially yourself) what people want, demand, and need in the short run. You can spend your whole career on the Low Ground and be very successful. But you need time on the High Ground to be the kind of leader who changes the world.

The strongest and most effective strategic leaders have trained themselves to consistently focus their attention on the High Ground. They know that they are choosing a more difficult way to live and work; it does not feel as natural. But they appreciate the results that follow. They gain influence and impact that goes beyond formal authority.

In this book, we lay out the difference between conventional transactional leadership and strategic leadership. We will show that strategic leaders focus not on what people want (the Low Ground in the mind and brain), but on the High Ground: "what needs to happen." We'll provide a four-step process for gaining proficiency in this—not just as an individual, but as a leader of organizations and community groups. We'll show how to deepen the inner voice of leadership, not just as a pattern of mental activity, but as a circuit "wired" into your brain. The leadership goal is to become a kind of Wise Advocate yourself, playing the same role in those larger organizations that this construct plays in your own consciousness.

Let's look at two real examples. The first story describes an extraordinary business challenge that threatened the integrity of one of the world's largest industrial companies. The second describes a professional woman dealing with setbacks at work that seemed like they would curtail her career.

Reframing Russian Reality

In the mid-2000s, William ("Bill") O'Rourke took two struggling, corrupt aluminum plants in Russia and turned them into models of safety and accountability. When he agreed to do this, he was a senior executive at Alcoa, the world's third-largest producer of aluminum, based in Pittsburgh and New York City.

O'Rourke had worked mostly on the staff side at Alcoa, in its legal, finance, procurement, auditing, and environment, health and safety departments. Then, in 2005, the CEO tapped him to take over its Russian enterprise, which largely consisted of two huge aluminum manufacturing plants. The company had acquired those facilities a year before to produce beverage cans and aircraft hulls for Russia's growing market. The assignment: make both plants profitable and bring them in line with global standards for financial performance, environmental quality, and employee safety.

Many experienced plant managers would have turned the job down. Both facilities were nightmares. They were rife with cost over-runs and mismanagement; environmentally, they were toxic swamps. Their safety records were so poor that the staff had stopped counting fatalities; serious accidents occurred continually, and one had resulted in the death of a plant manager. Corruption was also taken for granted; the head of human resources would later be caught skimming severance pay. The local governments on which the plant depended for security and transportation were even worse. When O'Rourke showed up, one city official welcomed him with a lightly veiled death threat: "If this was five years ago, I would kill you, and I would get away with it."

Though O'Rourke had never run a plant before, he'd spent thirty years at Alcoa, where he was known as a skilled "fixer"—a solver of difficult problems. He also had a reputation for unimpeachable integrity and discretion. All of that would be needed, because to bring the plants into line, he couldn't just rely on his authority. When it came to extortion, carelessness, pollution, industrial accidents, and scandal, the staff continually voiced the same deceptive organizational message: *These realities are more powerful than we are. It's not worth the effort to try to overcome them.* O'Rourke didn't believe this message was true, but he knew it was a self-fulfilling prophecy; the more credence people gave it, the more they would acquiesce, and the more daunting the problems would be. O'Rourke knew that his only chance for success was to spark a change in the way that his employees saw their world.

O'Rourke's first big opportunity came just a few weeks after he arrived. He had ordered a $25 million furnace for one of the two plants. It was located at Belaya Kalitva, about five hundred miles southeast of Moscow. The furnace, which had to be shipped from Germany, was essential; without it, they couldn't make aerospace plate, the material used to form airplane hulls, and without that market, Alcoa's operations in Russia would probably fail.

Local government officials demanded a $25,000 payoff to let the furnace enter the city. Bribes of this sort are like tests of resolve, and Alcoa had a long-standing firmwide policy against them; once you give in, your company becomes known as an easy mark. Even so, a message came to O'Rourke from Alcoa headquarters: he had to get the furnace installed and working no matter what it took, or his bonus would suffer. The Russian managers at the plant, who were used to paying bribes of this sort, advised him to negotiate it down to $10,000, which was small enough to be hidden in the budget.

Instead, O'Rourke decided to stand his ground and reject the bribe. "The furnace can rust on the street," he said. He knew he could not sustain this position by himself, so he began to systematically raise the awareness of everyone who worked at the plant. He pointed out that the conventional wisdom about bribery and industrial accidents in Russia—that they were inescapable—was just conventional wisdom, just a story they told each other. It wasn't necessarily accurate. (As we'll see later, this approach is an example of *relabeling*, the first of four steps of strategic leadership described in chapter 4 of this book.)

O'Rourke suggested a different way of interpreting events. He said Belaya Kalitva could become a major aerospace manufacturing city, a global economic hub—but only if Alcoa got the furnace and could put it to use. He said he understood why the local officials wanted the payment, but they should think of the furnace instead as an investment in the region's future. (This is the second step of strategic leadership; in chapter 5, we'll call it *reframing*.)

After only seventy-two hours of argument, the police let the trucks go through. They understood that Alcoa would not be susceptible to bullying or bribes. Beyond that, Alcoa would play a role in revitalizing the region if they let it.

Over the next few years, O'Rourke continued to stand his ground on bribes and corruption, both at Belaya Kalitva and at its sister plant in the city of Samara. When a tax inspector announced a surprise audit, avoidable with a special payment of $5,000 to the inspector, O'Rourke publicly welcomed the audit. When the import-export office refused to release goods unless the office manager received a direct fee, Alcoa refused to pay. When the supply chain slowed down as a result of refusal to pay some bribes, O'Rourke let everyone know the reason. (Exercising this type of consistency, paying attention to the wise alternatives time and time again, is a particularly powerful way to invoke the High Ground. It is the third step of strategic leadership, *refocusing*, which we will consider it at greater length in chapter 6.)

Finally, O'Rourke set up the kinds of practices and structures that reinforced his new messages. He started with the domain that affected people in the plant most directly: safety. If they are not designed and managed carefully, factories can be very dangerous places to work, and these two Russian plants had ten times Alcoa's global average scores for industrial accidents. It wasn't that people didn't care about safety; they just didn't have any experience with the regimens and thinking associated with it. They didn't follow the rules; they didn't wear safety equipment.

During 2006, his first year on the job, O'Rourke set a target of "zero incidents." No one else in the plant thought it was possible, but he insisted that they use it as a goal. It thus became a focal point in the plant's collective thinking. O'Rourke arranged for seven thousand people to be trained in safety practices. He made sure Alcoa provided protective equipment and enlisted the plant employees in finding and fixing fatality risks. They caught more than four thousand that year, and for the first time in living memory, there were no fatalities. Though he didn't call it by our name, O'Rourke was engaging in the fourth step of strategic leadership, called *revaluing* in chapter 7. He was creating a context, or a way of life, around this new, wiser way of looking at the world. Once they had achieved some success, O'Rourke began to remind people regularly: if they could change their thinking about safety, maybe they could change it about corruption as well. This, he knew, would be a more prodigious task. In that part of Russia, he later recalled:

> The prevailing attitude is that you are an idiot if you don't make money for yourself when you can. People expect to be extorted to get good medical care, to get their child accepted into a first-class

university, or to get any legal document processed quickly. When Russians take a driving test, they often have to slip a few rubles to the instructor. But I made our position on bribery and corruption clear: "We don't condone it. We don't participate in it. We are not going to do it. Period." I was realistic, but I never judged the culture as irredeemable. America 100 years ago was not unlike Russia today. Ethical maturity takes time.[11]

Gradually at first, and then with more speed, the culture of Alcoa's business in Russia shifted. Both plants went more than seven years without a fatality; their safety record was better than Alcoa's global average, and they were approaching that impossible goal of zero incidents. In 2011, O'Rourke retired; he is now a lecturer and author on business ethics and a fellow at the Wheatley Institute at Brigham Young University.[12] Alcoa Russia, which had sixty-two expatriates from eight countries on its 2005 management team, is now managed solely by Russian citizens; it still maintains an ethic of openness, safety, quality, and accountability. As of 2017, it is the largest producer of fabricated aluminum in Russia. This is in a country where many other multinationals have lost control of their assets or departed, and where political conditions and sanctions can make business difficult. The company sold the Belaya Kalitva plant in 2015 to a Russian firm, Stupino Titanium; it continues to operate the Samara plant, which continues to deliver solid performance in aerospace technology.[13]

The point of this story is not that O'Rourke stood his ground against significant pressure from many stakeholder groups. Many leaders do that, just to hold the positions they have. O'Rourke did something subtler, but more significant. He created the conditions under which people changed their thinking in a way that outlasted the time he spent there.

An HR Executive's Dilemma

Another executive we know who became a strategic leader by using the High Ground is the human resources director for a major professional services organization, a linchpin in its local economy. (We have permission to tell this story, but we cannot use the real name of the company or the individual.) "Natalie," who is in her forties, reported directly to the CEO. When the firm hit a long stretch of dwindling revenues, Natalie had ideas for turning things around, but she wasn't included in strategic conversations. Instead, all personnel issues—including sexual harassment cases, bullying claims, and layoffs—were delegated to her. One year, she had to move the firm's financial accounting staff offshore. About thirty local people lost their jobs. It was a painful but necessary decision that allowed the firm to survive. Stress took its toll. For years, Natalie worked seventy hours or more per week. Her marriage was on the rocks, she came to work anxious, and she lost the ability to hide her chronic irritation. As a result, her performance reviews slipped. She felt herself panicking: *If this goes on much longer, I won't be able to cope, and I'm going to lose my job.*

Fortunately for Natalie, she knew people who gave her some outside perspective, helping her see what was happening, and encouraged her to think that way herself. First haltingly, and then with growing enthusiasm, she adopted a regimen of practices that included mindfulness.

Every day, soon after arising, she spent a half-hour alone, focusing her attention on the deceptive brain messages that underlay her stress. For instance, she knew she tended to see everyone but herself as prone to error. *Most people are screw-ups, and need to be tightly managed.* She also believed that the firm's leaders didn't respect her. *I'm just the head of HR, and the real work happens in sales and finance.* She used to assume these were accurate statements of reality, and they led to many of her worst decisions. She resented the leadership of the organization and didn't even try to find a way to voice her thoughts. She micromanaged her staff because she felt they could not be trusted. Of course, she could not keep the resentment, of people higher and lower in the hierarchy, from showing.

But now she has *relabeled* these assumptions as brain messages—not unlike the way Bill O'Rourke relabeled the messages about bribery being a natural element of doing business in Russia. Natalie can observe dispassionately as these messages rise into her awareness. Deceptive brain messages like these are generated by habitual, near-automatic brain processes. They are not necessarily neurotic; everyone has constant experience with them. But they are always, at best, incomplete depictions of reality that seem as though they're complete. Thus, although they are often inaccurate, they take on a power within the mind; if they are not examined, they can lead people to make poor decisions.

As Natalie reflected, she *reframed* her thoughts, choosing alternative ways of looking at her situation. These didn't come out of thin air; she practiced thinking through the firm's problems—sometimes in areas she knew well, such as recruiting and training, but also in less familiar domains, such as mergers and growth—and proposing strategic approaches. She also looked closely at her belief that the major decision makers didn't respect her. Where had that belief come from? Why did she tend to ruminate on these messages of insecurity? She concluded that a dispassionate observer would not see chronic disrespect—indifference at times, maybe; but she could manage indifference.

At the same time, Natalie began systematically thinking through reasons why the firm's business might be declining and where it might realistically grow, given its own potential. The conclusions weren't always obvious, but once she looked, she saw possible ways to improve the company's prospects, particularly through means she knew something about: recruiting, training, designing incentives, and sharing knowledge. Through this process of reframing, she came up with useful, true alternative messages. *I know what this company needs to do, and I can make the case for it.*

With these alternative messages in place, she *refocused* her attention on them, returning again and again, for example, to ways in which she could make a valuable contribution. Consciously, willfully, she devoted her attention to them—thinking about them repeatedly, deliberately rehearsing the new thoughts until they felt natural to her. Before any major meeting, she thought about how the various leaders of the company might respond to the points she would make. As she made critical decisions, she reminded herself to pay attention to the way others responded and followed up. She looked at how the top team considered her insights and how the people reporting to her followed up. She observed, without judgment, her own ability to observe and how it was improving.

Natalie began this discipline around 2013, and it gradually affected the way she spoke and the things she said. She is now regularly invited into conversations about strategy. When there is a possible crisis, people turn to her first, as if she were a Wise Advocate for the larger enterprise. She identifies and brings deceptive brain messages to the surface—*this is a false sense of panic; we actually have until next week to make this decision*—and proposes alternatives, inviting others to share her growth and opportunity-oriented mind-set. The company's prospects have turned around—in part because of opportunities she has pointed out—and instead of laying people off, she's now recruiting. She has also reduced the amount of oversight and number of approvals in the HR function; she no longer has to work seventy hours per week. She has, in effect, *revalued* her relationship with the organization and can thus operate at a recognizably higher level of credibility, influence, and impact.

You might think this is just standard good management practice, nothing special. And you may well be right. But it was beyond Natalie's skill four years ago. She made a deliberate transition, from a harassed functionary bent on pleasing her bosses to an influential leader with strategic perspective. The potential for this change was there all along, but nothing external—no incentives, rewards, threats, or "burning platform" pressure—could force her into it. The leverage came from transforming her thoughts. By relabeling, reframing, refocusing, and revaluing, she became the kind of leader needed in that company at that time.

The shift that Natalie made was conscious, pragmatic, and replicable; anyone reading this can make it too. The same is true of the stand that Bill O'Rourke took at Alcoa. These stories represent two of many examples of our hypothesis about leadership—that better, more strategic leadership can be developed by combining two often-misunderstood cognitive habits: mindfulness (clear-minded awareness of one's own mental activity) and mentalizing (paying close attention to what other people are thinking and are likely to do next). For all its complexity, the wise leadership hypothesis, as we sometimes call it, can be boiled down to one core principle: *The focus of your attention in critical moments of choice can build your capacity to be an effective leader.*

You might interpret Natalie's story as saying, "The solution was there all along. She was in charge all the time, but just didn't realize it." That's not what we're saying. There were real forces all around her that she had no control over, that she was powerless to stop. But she had leverage in her thoughts. By reframing the messages and refocusing her attention, she could bring the activity in her mind from Low Ground to High Ground—from a focus on expedience, self-referencing/self-affirming, and pleasing people to the perspective of a Wise Advocate. She's still not in control of everything; but she no longer feels she needs to be.

Leaders on High Ground

When leadership of this sort occurs, it tends to be remembered. No one in South Africa can forget the example of Nelson Mandela who, when released in 1990 after twenty-seven years in prison, invoked the High Ground by championing a democratic government and the truth-in-reconciliation process that allowed the country to make a peaceful transition from the racist authoritarian regime it had been to a consciously inclusive democracy. One of the many examples of Mandela's High Road occurred in 1995, shortly after he was elected president of South Africa. Most of the other leaders of his party, the African National Congress, wanted to change the name of the national rugby team, the Springboks, which had long been associated with apartheid. Mandela insisted on keeping the name and wearing a Springbok jersey when the team was up for the World Cup, because he consciously knew that this symbolic gesture would show the white population of South Africa that they were included.

"I am proud to wear [the symbol]," he told a reporter, "when a few years ago, even a few months ago, it was anathema . . . this thing, that was a very divisive and ugly symbol, could in fact have been magically used by God to weld us together. No one of us could ever in their wildest dreams have been able to predict that rugby could have this magical effect."[14] (Since Mandela left office, most South African political leaders have been more closely associated with the Low Ground, with devastating effects on the country's prospects.)

Ed Whitacre, the CEO who brought General Motors out of bankruptcy in 2008, had some of that same High Ground aspect. Intent on raising the quality of GM's cars, he was the first GM president to meet with the head of GM's union at the union headquarters, rather than the company's executive floor or a neutral area. This small gesture was one of many that he took to show that he was serious and that the company could not afford the status-driven rivalries, beset with deceptive messages, that had shaped its management style in the past.[15]

Mark Bertolini, as we'll see later in the book, did something similar as CEO of Aetna. So did Frances Hesselbein when, as CEO of the Girl Scouts, she focused on fostering skills related to science and technology education; Howard Schultz at Starbucks when he curtailed growth plans until he could raise the quality of his practices; and

Soichiro Honda at the automobile company that bears his name, when he developed day-to-day practices for learning from failure throughout the company.[16]

In the political sphere, it's harder to name specific people who everyone would agree exhibit High Ground leadership. But there are a few people who consistently seem to avoid being caught up by events, who know how to see themselves as others see them. You might not agree with their policies, or the way they play the political game, but you can generally see that they are not just trying to fulfill their personal and professional appetites; they seemed to be in it for something deeper, bigger, and more important than their own status or power, and they exercise that ambition with a high level of self-awareness.

It's probably easier to identify them in your own organization: people acting on behalf of the true identity of the enterprise as they see it, even when it isn't expedient. We think there are many potential and actual strategic leaders in the world today, including many members of this book's audience. Most of these individuals have deliberately cultivated this quality in themselves.

Seeking Your Own High Ground

If you are interested in developing this type of leadership in yourself, then we have written this book for you. Its approach may seem "soft" or "unsubstantial" at first glance. We're suggesting that you can influence the world around you by changing the ways you place your attention, as opposed to with persuasive rhetoric, positional authority, or force. But the application of attention is the most practical tool you have. It gives you leverage over every material-seeming structure in your life, from the biology of the brain to the systems and technologies of a vast company.

Focusing your attention in a High Ground manner may seem difficult at first, and it does take persistent, recurring effort. (That's why we call part of this process *re*focusing.) Attention is like water—it tends to flow downhill, on the path of least resistance, to whatever is most salient (salience is the capacity to attract attention, generally by being obvious and distracting). When your attention as a leader is left to itself, you will naturally find yourself on Low Ground. That's the most direct route to getting what you want, and to doing what others want.

At an individual level, moving to the High Ground—and the Higher Ground beyond it—will engage you, day after day, in a life-long process of discovering your "true self": the identity you seek to realize, because it is closest to who you want to be in spirit. Every individual has a true self, a core of personal belief and identity, and powerful leaders tend to have spent time getting in touch with it. This, too, involves returning your attention to your Wise Advocate: the part of your mind that sees your true self as worthwhile, wants you to move closer to it, and has a clear, wholesome sense of what direction will take you there.[17] It's easy for individuals to get distracted and lose touch with their true selves. That's why, if you hope to be effective as an individual, you need to consciously cultivate a Wise Advocate within your own mind.

If all this sounds a bit self-referential—your "true self" is the source of your Wise Advocate, and your Wise Advocate will take you to your "true self"—that's because we have no other way to write about it in a general piece of nonfiction. Your true self can't be completely articulated by a book. It is not universal. Its nature is specific to your own life, your goals, your relationships, and your long-term aspirations. The practice of the four steps of strategic leadership—relabeling, reframing, refocusing, and revaluing—can significantly help clarify and articulate it. Indeed, many very old paths of deeper knowledge, paths that have been practiced for millennia, include some form of the practices described in this book. They have helped people throughout human history improve their actions and gain a more enlightened life.

We believe that every organization also has a "true self"—some core identity that affects everything it does. With a company like Apple or IKEA, for example, you know why it exists, and you would recognize a gap, or loss, if the organization went away. It's even easier for organizations to lose their way—to get out of touch with their collective true selves—than it is for individuals. After all, the organization is composed of diverse individuals, all with their own concerns and agendas, often with conflicting aspirations and perspectives. The leaders of the organization may or may not even recognize the extent to which strategies and capabilities are aligned. They may or may not value a long-term outcome; they may have their attention focused on short-term profits and expedient measures. They are always potentially vulnerable to deceptive messages. The stakes are high in any major organization, and the presence of one or more Wise Advocates may make all the difference.

Bringing the Ideas Together

The rest of this book is organized to explain these concepts in more detail, along with the neuroscience and organizational research that illuminates them and the ways in which you can adopt and deploy them effectively. The next two chapters focus on the High Ground, the Low Ground, and related concepts, including more explicit descriptions of the mental activities associated with these two paths.

Chapter 2, "Low and High Ground," covers the two main patterns of mental activity that you might call upon in a decision-making moment, and the brain circuitry related to them. We also explain the nature of self-directed neuroplasticity—the fact that focusing your attention and moving your mind more frequently to High Ground activity enhances your brain function.

Chapter 3, "Finding Your Inner Voice," discusses the types of mental activity that can invoke the Wise Advocate within you: mentalizing (thinking about what other people are thinking and what they are going to do); executive function (which includes regulating your impulses); and several forms of mindfulness (including the skills of meta-attention, meta-cognition, and meta-awareness). It also describes your "free won't"—the innate ability that people have to recognize brain-based impulses and cravings without giving in to them.

The remaining four chapters help you bring these concepts into your own leadership practice. They each describe one of the four critical steps of strategic leadership that you can take to move yourself to High Ground and closer to the Higher Ground.

Chapter 4, "Relabeling Your Messages," talks about the deceptive messages rampant in organizations and how leaders can bring them to awareness and stop their devastating impact. It contains a taxonomy of common deceptive messages that distract many transactional leaders.

Chapter 5, "Reframing Your Situation," describes how to replace deceptive messages with your own chosen messages that can help you move in new directions. With an explanation of the placebo effect, it shows how channeling your expectations can help make a reframed message more compelling and powerful.

Chapter 6, "Refocusing Your Attention," describes the ways in which sustained attention changes entrenched patterns of behavior in your brain, in your personal life, and in the organizations around you. It lays out the discipline that makes genuine behavior change possible and the mental activities involved in it: self-directed neuroplasticity, attention density, and the quantum zeno effect.

Chapter 7, "Revaluing Your Leadership," discusses the ways in which your High Ground activities set a context for new, more powerful strategic leadership. It discusses some of the dynamics, such as the organization's core group, that determine whether your Wise Advocate ideas are likely to be seen as legitimate.

The three of us—a cognitive scientist, a journalist on the business strategy beat, and an executive coach—have been talking with one another about the concepts in this book since the mid-2000s. But we could not have written it until now. It is only in the last few years that the view expressed here of the interface between brain mechanisms and behavioral change has emerged to a great enough extent that we can see the High Ground for what it is.

On the neuroscience side, there have been breakthroughs in imaging technologies such as functional magnetic resonance imaging (fMRI) and positron emission tomography (PET), along with brainwave analysis technologies such as quantitative electroencephalography (QEEG). These have revealed hitherto unseen neural connections in the living human brain. Advanced computer analysis of these connections has helped psychological researchers develop an increasing body of theoretical work linking the brain (the physical organ) with the mind (the human consciousness that makes choices and decisions about what it thinks, feels, does, and perceives).

On the organizational side, theories of management and leadership are evolving toward an improved understanding of how people work together effectively and how to promote effective change. Though group dynamics and organizational learning have long-standing histories, it is only recently that people are combining those with the insights of neuroscience.

Finally, on the ethical side, there is an increasing recognition of the importance of strategic leadership, and especially of the inner voice. More and more people understand that when leaders are abusive, the entire organization is responsible—at least for speaking out about it. They also understand that organizations, whether private, government-based, or not-for-profit, cannot survive intact if they persist in the same practices and carry on with the same ethical values, unquestioned.

"That's the way we've always done things around here" does not suffice as an excuse.

The deep dialogue about ethical values that has taken place throughout human history—seeking to understand the purpose of life and the effect this has on day-to-day choices—is thus moving into the arenas of neuroscience, psychology, and management. A growing number of leaders in these fields are beginning to recognize that any worthwhile discussion about change in human activity must consider these questions: Who do you want to become? What values are driving you to change? What effect is this change having in reality? Is it helping you reach your fundamental goals, or are you perhaps falling for a deceptive message? How can you know the difference?

We draw on much of this research, as well as on some of our own writing on neuroscience and leadership.[18] We also draw heavily on *You Are Not Your Brain*, a personal psychology book coauthored by Jeffrey Schwartz and psychiatrist Rebecca Gladding, in which the four steps of strategic leadership, along with the terms Wise Advocate and True Self, were introduced.[19] As a coach, Josie Thomson has been working with organizational leaders using the same relabeling, reframing, and refocusing practices described in that book and has developed a track record helping teams and companies move to new levels of success. Art Kleiner's writing about organizational learning and change, particularly in *The Fifth Discipline Fieldbook* and *Who Really Matters: The Core Group Guide to Power, Privilege, and Success*, has provided an important complementary perspective, particularly around the relationship between leaders and their organizations.[20] We resolved to write a practical guide for leaders seeking better impact and influence, toward beneficial ends, that tried to make sense of all these threads together.

Many people in business already know something about neuroplasticity—the innate tendency of the brain to change physically in response to a change of focus in attention. Many are practitioners of mindfulness—the capacity for making clear-minded observations of one's own inner mental life. But few know how to use all this as a leader, in a systematic and consistently productive way. Even the experts in the field are still figuring out how all the pieces fit together. There are guides, but they are often incomplete, ambiguous, difficult to follow, or even misguided. We have written this book to fill the gap. We hope to help you become a kind of Wise Advocate yourself, playing the same role in those larger organizations that the internal Wise Advocate plays in your own mind.

Reflection

Think about your current decision-making style. How would you describe it—expedient, seeking rapid relief, or more thoughtful and discerning? What happens at that moment of choice?

If an answer doesn't come to mind, think about one or more recent major decisions you have made. What was involved?

How would others describe your typical decision-making approach? Would they characterize it the same way you do?

How long does it typically take you to decide on a course of action on a difficult matter?

Where do you turn for guidance? Do you have a personal moral compass? A person or group you consult? A view of what you want, or what others want? A considered approach for discerning viable options efficiently and effectively? An internal or external Wise Advocate?

Do you know your true self? What are your core values and guiding principles as a leader?

CHAPTER TWO

Low and High Ground

WHEN HE WAS still in high school in the mid-1980s, the venture capitalist Reid Hoffman—one of the cofounders of LinkedIn, the most prominent online network for professionals—decided that there were four types of people in the world. The first category is purely transactional. People of this sort, when considering a deal or alliance, want to know immediately what the payoff will be before they agree. "I'll do something for you," they insist, "only if you will do something for me."

The second group is only slightly less transactional. The payoff doesn't have to be explicit, but it does have to be equivalent. "I'll do something for you, but I'm keeping track of what you owe me." A third group is transactional with a longer time frame: "I'll invest in this relationship, and I'll expect you to invest commensurately over time."

Hoffman says he has learned to make deals with individuals in all three transactional groups, and to appreciate them, but he never invites them to work closely with him. He knows that when push comes to shove, they will choose expedience—the ability to get a certain short-term gain—over any longer-term benefits or opportunities, even if those are much more desirable in the long run.

But there is a fourth group: people who are not transactional at all. "I'll do something for you," they might say, "because I believe in my relationship with you, and making this investment is the right thing to do." Throughout his career, Hoffman has picked his closest associates—who often become very wealthy through their association with him—from this group. He sees them as following a nontransactional approach to life. They don't necessarily require their allies and counterparts to make an explicit commitment for a short-term payback, but they tend to prosper more than the transactional group in the long run.

One example of the fourth group, of course, is Hoffman himself. In an article he wrote about this way of categorizing people, he told a story about a million-dollar deal he made in the mid-2000s. He and his longtime associate Mark Pincus, CEO of the game company Zynga, had worked closely together on funding a series of social-network patents, and they had developed an understanding that they would share all opportunities; if one stood to prosper, so would the other. Then Hoffman was invited to make an early investment in Facebook, long before its initial public offering, and Pincus was not. This was clearly one of the most valuable investment opportunities imaginable, worth many millions of dollars.

"I realized," Hoffman recalled, "that given my relationship with Mark, the right thing to do was to give him the option of taking half of [my investment opportunity]." If Hoffman didn't make that offer, then they might find their interests diverging on their other investments. Nor did Hoffman ask Pincus for something in return, or for a guarantee that he would do the same thing by a specified time. He made the offer to Pincus without strings. They continued to work together thereafter, and their alliance has greatly influenced the evolution of the online world.[1]

The first three categories of people are used to thinking in terms of negotiations. If you give people what they want, they will give you more of what you want, either explicitly or in a broader way. Knowing what they want will help you strike a better deal. If you compete with them, it will help you win. There's nothing intrinsically wrong with this approach; but it keeps you on Low Ground, following the pattern of mental activity based on expedience.

Hoffman, on the other hand, seems to gravitate to the High Ground. While he pays some attention to what people want, he puts more of his attention on what they are thinking—and what they are likely to do. He is skilled at this, and for good reason: he's been studying people in this way since he was a teenager. Indeed, by the time he came up with his typology, Hoffman had already decided that he was going to build his career around business relationships. (*Wired* magazine has called him "the best-networked man in Silicon Valley."[2]) Focusing on what people are thinking and what they are likely to do involves looking at them dispassionately—not in terms of competing with them, striking a deal with them, or pleasing them, but in terms of who they are and how they affect the system around them.

To be this kind of nontransactional player, Hoffman also has to think about himself in an unusual way. He has to become intensely self-aware—aware of his own thoughts, emotions, and motivations in a similarly dispassionate way, able to simply observe himself. This is less evident in his own writing, but it is a perennial theme in what people who work with him have written about him. For example, his former chief of staff Ben Casnocha, who coauthored two books with him, wrote:

> [Reid's] ability to manage his "alpha streaks" [to rein in his reliance on the traditional markers of status and power] partly explains why his partnership with Jeff Weiner at LinkedIn has worked well. Reid was able to acknowledge he needed to hire a CEO to replace himself, which is not a realization every founder can come to terms with. Jeff was comfortable having a smart, influential founder still active at the company as executive chairman, which is not a dynamic every CEO can stomach.[3]

This type of long-term thinking is typical of the High Ground. It allows Hoffman to distinguish among deal partners in a more productive way than simply asking "Who has better financial prospects?" He can still evaluate the financials, but the most critical factor is his prospective partners' recognition of "the right thing to do" and their willingness to set up their businesses and investments accordingly. To establish and maintain a fruitful relationship with this type of individual, Hoffman too has to be willing to make choices based on his long-term goals—and he has to follow through. When you're operating from a stance of integrity, too much is at stake to let a deal slip through the cracks by being expedient.

Patterns and Circuits

Figure 2.1 shows how the High Ground and the Low Ground relate to each other in both the mind and the brain. Each is represented by a loop: High Ground above Low Ground. Each loop connects three primary centers of mental activity, represented by the circles. Within each circle, you'll see:

- A bold name (the Habit, Warning, Executive, and Self-Referencing centers) identifying its essential function in both brain and mind
- In smaller type, a name drawn from neuroscience for the brain area associated with that function—the area where researchers consistently find neural activity related to it (lateral prefrontal cortex, basal ganglia, and so on)
- Another name below that, this time drawn from psychological research, for the mental activity associated with that center (mentalizing, subjective valuation, and so on)
- A quote in italics representing the related mental activity as you might experience it (*"Something is wrong"* and so on)

Each loop thus shows the mind and brain together. With regard to the mind, each loop represents a pattern of mental activity, incorporating the experienced thoughts and messages from each of the functions involved. For the brain, each loop represents the physical circuits associated with those patterns, sending and receiving neural signals across the related regions of the brain.

In this book, when we refer to a Low Ground pattern or a High Ground pattern—or just Low Ground and High Ground—we are referring to the recurring patterns of mental activity in the mind. When we refer to a Low Ground brain circuit or a High Ground brain circuit, we are referring to the physical counterpart, the neural circuit in the brain. You'll find that we have written far more detailed descriptions of the patterns of mental activity than of the brain circuits; that's because, while the brain circuits help explain what is going on and why the Low Ground and the High Ground have such an impact on your effectiveness as a leader, there is a more relevant, deeper understanding—and far more leverage for action and leadership development—in the patterns of mental activity.

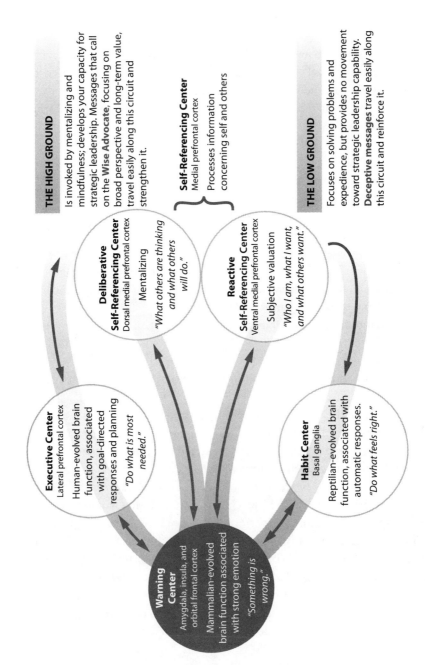

THE HIGH GROUND

Is invoked by mentalizing and mindfulness; develops your capacity for strategic leadership. Messages that call on the **Wise Advocate**, focusing on broad perspective and long-term value, travel easily along this circuit and strengthen it.

Self-Referencing Center
Medial prefrontal cortex

Processes information concerning self and others

Deliberative Self-Referencing Center
Dorsal medial prefrontal cortex

Mentalizing

"What others are thinking and what others will do."

Reactive Self-Referencing Center
Ventral medial prefrontal cortex

Subjective valuation

"Who I am, what I want, and what others want."

Executive Center
Lateral prefrontal cortex

Human-evolved brain function, associated with goal-directed responses and planning

"Do what is most needed."

Warning Center
Amygdala, insula, and orbital frontal cortex

Mammalian-evolved brain function associated with strong emotion

"Something is wrong."

Habit Center
Basal ganglia

Reptilian-evolved brain function, associated with automatic responses.

"Do what feels right."

THE LOW GROUND

Focuses on solving problems and expediency, but provides no movement toward strategic leadership capability. **Deceptive messages** travel easily along this circuit and reinforce it.

Figure 2.1 High Ground and Low Ground. *Source:* Jeffrey Schwartz, Josie Thomson, Art Kleiner, and Wise Advocate Enterprises.

Let's look at these two patterns separately first. Then we'll consider the means by which you choose between them at the moment of decision and how this aspect of human learning results in brain change, or neuroplasticity. Because of neuroplasticity, the more attention you pay to either of these patterns, the stronger the brain circuits associated with it become, making it feel easier and more natural. We'll then show how neuroplasticity can be channeled—or, more precisely, self-directed—to help you shift more and more easily from Low to High Ground, first in your mind, and then with more habitual, automatic responses in your brain.

Running on the Low Ground

Life today is a constant barrage of challenges. You have promises to fulfill, problems to solve, tests to pass, and situations to manage. The Low Ground represents the pattern of mental activity involved in meeting these challenges in an expedient way. When you make deals, design rewards and incentives, or think about satisfying your needs or the needs of others in your organization, you tend to place your attention on the Low Ground. This activity often elicits powerful emotions, such as desire, anxiety, fear, frustration, elation, and relief. In everyday workplace life, most of us occupy the Low Ground most of the time.

Low Ground activity is also known as "subjective valuation" because it is concerned with what is valuable and relevant. *What's in it for me? How much is it worth? How might we close the deal? What might others want?* Though powerfully related to incentives of various kinds, these are not purely selfish concerns. This is the pattern in your mind that asks *What's in it for me?* but also asks *What's in it for the other person?* Any thoughts of reward, for yourself or others, trigger activity here. For example, if people you know receive a prize or a grant and you are happy for them and take pleasure in their success, that's a Low Ground response. (So is feeling envious of them.) Low Ground mental activities align well with loyalty, at least to the degree that giving people what they want keeps them loyal.

The Low Ground appears to be the primary mental pattern of transactional leadership. It involves solving problems—figuring out what will make you happy, or someone else you care about happy, and doing whatever is necessary to relieve the tension and discomfort of dealing with an unsolved issue. Chances are, when you're active in this pattern, you hope to reach your goal as quickly and easily as possible. That's why the word "expedient" defines Low Ground decision making. People who follow this pattern tend to look for simple solutions where the rewards are immediate and the relationship between cause and effect is easy to understand.

This circuit connects three major functional centers, each associated with a region of the brain and a particular type of mental activity. We call the first the Reactive Self-Referencing Center; it is associated with the ventral medial prefrontal cortex (vmPFC). This center is spontaneously activated when there are thought processes or sensory stimuli perceived as primarily related to the self.

It is important to note that the Reactive Self-Referencing Center is just half of a larger system called the Self-Referencing Center. As we'll see, the other half (the Deliberative Self-Referencing Center) is a key element of the High Ground. The overall Self-Referencing Center is far more developed in human beings than in any other animal. This center is associated with the medial prefrontal cortex, which is involved in many aspects of your personality and identity—especially in the way you perceive yourself and relate to others.[4] It correlates with your inner monologue—the voice inside your mind that articulates your hopes and fears, daydreams about the future, thinks about people, and interprets experience. When you're on the Low Ground, this inner monologue will be oriented to yourself; as we'll see later, it's different on the High Ground, which is much less prone to subjective valuation.

A second function of the Low Ground is the Warning Center. It is associated with three parts of the brain: the amygdala, the insula, and the orbital frontal cortex. This center generates feelings of fear, gut-level responses, and the sense that something is worth pursuing or avoiding. Anxious feelings of impending danger (especially those related to the experience of past threats) can activate this center with such intensity that they override all other thinking and response. With the publication of *Emotional Intelligence* by Daniel Goleman, this phenomenon became known as the "amygdala hijack."[5] (As we'll see, the Warning Center is also associated with the High Ground.)

The third major brain function on the Low Ground is the Habit Center. This function, typically associated with the basal ganglia (which are located deep within the base of the brain), manifested itself early in animal evolution. (It is sometimes called the "lizard brain.") Habits, in this context, are automatic thoughts and actions—basic behaviors that don't generally require conscious attention because they have become automatic through repetition. These are actions such as walking upstairs, locking the door, brushing your teeth, and steering your car. The Habit Center manages them. Making better use of this center by retraining your automatic habits is the subject of Charles Duhigg's bestseller *The Power of Habit.*[6]

If you have pleasurable sensations associated with the Low Ground (and most people do), it's because among the many components of the basal ganglia is a part of the brain called the accumbens. The accumbens, which is related to the Habit Center, is also called the reward center, because it is a major source of pleasurable feelings; it is often triggered by sex and addictive substances. Thus, even if the Low Ground is problematic for you—even if you become aware that there are deceptive messages traveling on it, leading you and your organization to overly complacent, counterproductive, addictive, or self-defeating behavior—you may find yourself and your colleagues resisting any effort to change. That's because neither you nor the rest of the people in your organization want to lose the appealing, powerful mental sensations associated with your old collective habits.

In *You Are Not Your Brain*, the Low Ground is called the "It's About Me" circuit.[7] In this book, we have discarded that name, because the Low Ground is not *just* about you. It is not necessarily selfish. Rather, it is oriented toward the short-term rewards of worldly activity: prices, exchanges, and deals. This circuit is directly related, via the accumbens, to the pleasure you feel when getting an immediate reward, like winning the lottery, getting a present, or receiving a bonus or stock dividend payout. It is also, via the Reactive Self-Referencing Center, the place where you determine how much you want the reward in the first place—and how much you think others might want it. When you feel happy because someone else gets a reward, that is also a Low Ground feeling. So is the pride you may take in having something others want, or the admiration you may feel when others have something you think is valuable. In this pattern of mental activity, you'll often experience brain messages related to reward, gratification, and relief. When you ask yourself *What's in this situation for me?* or *How do I get what I want?* or *What do other people think this is worth?*, you are traveling on the Low Ground.

Because it is connected with subjective valuation, the Low Ground is often evoked by deal-making. When you think about winners and losers, rewards and punishments, and the financial outcome of a deal or a bargain, your mind is probably focused on the Low Ground. The idea that everyone has a price is a Low Ground concept, and the Low Ground is involved in figuring out how valuable that price is. When the price is right, activity in the Low Ground goes up. It may also go up whenever you think about economic exchange of any sort. Subjective value is closely linked with the founding of the Austrian school of economics, and it is thus linked to the idea that the worth of something depends on what others are willing to pay for it, not on, say, the experience of producing it or the creativity that went into it.

In recent years, a number of psychology researchers have studied functions that we consider relevant to the Low Ground. They have made some fascinating discoveries that seem to support Low Ground qualities. For example, Oscar Bartra of the University of Pennsylvania describes the vmPFC, or what we call the Reactive Self-Referencing Center, as a region associated with perceptions of prospective reward.[8] When things come to mind that you associate with a generally positive value, this brain circuitry is activated. Jamil Zaki of Stanford University has found that "value computation" is a hallmark of functions that we have associated with the Low Ground.[9] To Zaki, these functions are "person-invariant"—engaged just as much when thinking about a reward for someone else as it is when thinking about your own reward. Matthew Lieberman of UCLA, who has written a great deal on the cognition related to social interaction, has linked this type of cognition—again, which we associate with the Low Ground—to thinking about yourself and people you perceive as similar to yourself.[10]

You can see Low Ground dynamics in a number of familiar social situations. The maven (local expert) is a Low Ground creature. If someone says, "Stick with me, I can get you the best ice cream in the city," he or she is engaging in subjective valuation. The Marxist is also a denizen of the Low Ground. When used to assert a political agenda, the statement "From each according to his ability, to each according to his needs"[11] inevitably suggests that every worldly activity has a price, inevitably determined by the subjective valuations of some politically powerful interest group. Business success and failure also trigger thoughts of value and financial exchange. When people gain or lose confidence in a company's stock—for example, when they think

If only I had bought Apple stock in 1999—that's usually taking place in the Low Ground of their minds.

When you feel bonded to people, your efforts to help or reward them tend to trigger this circuit. Loyalty, especially when voiced in practical terms such as those of a favored transaction, is often a Low Ground emotion. When you want people to think well of you and you do something to help them—for example, when you solve a problem for your boss, your spouse, your coworkers, or your children—the Low Ground is typically engaged. It is also probably engaged if you're playing a game, particularly a serious game, in which there are winners and losers and a meaningful payoff or jackpot for the winners. This means that in many organizations, where measurements and rankings determine success, the Low Ground is continually in play.

Earlier, we said that the Low Ground is not necessarily bad for you or your organization. That's true—it can help you solve many problems rapidly and keep up with the flow of opportunity. For example, if a potential moneymaking opportunity catches your attention at a trade show, the Low Ground could be your source of urgency and follow-through. It can also provide you with insights that help you manage the constituents of your enterprise effectively. What would consumers pay for your product? What bonus will your employees accept? What does your boss want, right this second or in the next quarter? How high should you price your stock? Questions like these trigger the Low Ground, and your career may prosper if you answer them shrewdly.

Your grasp of subjective valuation can also make you passionate and persuasive to others, especially with messages related to their immediate goals and needs. Some gifted and charismatic (albeit narcissistic) leaders are extremely skilled at traveling the Low Ground. They can "read a room" and "give the people what they want," powerfully and decisively, and they thus come across as masterful competitors. The best of them are indeed great achievers, whose instincts are in sync with the needs of their time. And everyone who follows this circuit knows the feeling of competence and achievement that accompanies it. Former GE CEO Jack Welch titled one of his books *Straight from the Gut*,[12] invoking the idea that signals from this circuit can be powerful and tend to feel true.

But though they tend to feel true, these signals aren't necessarily accurate. Deceptive brain messages frequently arise from the Low

Ground, ranging from all-or-nothing thinking (*You're either a winner or a loser in this company*) to overconfident exceptionalism (*Our big customers have nowhere else to go*). As described in chapter 1, Natalie's chronic worries (*I will never be taken seriously as a leader of this enterprise*) were deceptive Low Ground messages. So are many other messages of expedience, including rationalizations for crossing an ethical line (*No one will notice if we manipulate these numbers*). Recall from chapter 1 the advice that Alcoa executive Bill O'Rourke received, to negotiate a bribe down to $10,000 so it would not be noticed on a balance sheet and would remain under the radar of accountability. That was a Low Ground solution to a profound moral problem. It's also the type of solution that could come back to haunt a company, because other government officials could hear about the bribe and expect as much or more next time. (We'll explore these deceptive messages in more detail in chapter 4.)

The automaticity of the Low Ground makes it dangerous: it leads people to feel that they should trust these brain messages and act accordingly, no matter how deceptive the messages may be. It is thus a source of inertia, a spur to continuing in the same direction that you're already taking. People who regularly invoke this circuit (which includes all of us, at least some of the time) get accustomed to thoughts, especially about themselves, that are not fully accurate. For example, in social situations, messages on the Low Ground might involve a sense of being misunderstood (*That's not fair!*) from your Self-Referencing Center, bolstered by the Warning Center (*People are going to believe this lie! What's going to happen to me?*). Or you might find yourself becoming complacent (*We're all right as we are; we don't have to worry about that new competitor*). Perhaps the most disconcerting aspect of the Low Ground is that it can readily become almost entirely controlled by the Habit Center, so that your decision making as a leader essentially runs on automatic. (In chapter 5, we'll look at ways to reframe these Low Ground messages, and thus reduce their influence on you and your organization.)

The field of system thinking, one of the more farsighted management disciplines we know of, has a name for one type of Low Ground solution: "shifting the burden." In this dynamic, which occurs in just about every type of system from an ecological environment to a school system to an addiction problem, there is a problem calling for resolution. Two different solutions present themselves. One is a quick fix, while the other takes longer to show results. People know the dangers of the quick fix, but they also find it irresistible. Thus, for example, educators may feel forced to adopt high-stakes standardized tests, even though those do not provide a clear picture of students' real learning achievements, because it is too slow and difficult to establish more thoughtful ways of evaluating students. Corporate leaders may slash costs across the board, eviscerating both valuable and problematic programs, because they fear it will be impossible to truly evaluate each expense on its merits. Alcoholics will take one more drink, because they know that it will provide rapid relief. In all these cases, the quick fix only partially solves the problem; but because the burden of the problem is shifted onto the quick fix, the participants believe they are closer to a solution than they actually are. This gives the problem an opportunity to grow worse. It's a safe bet that when there is a shifting-the-burden dynamic at play, the Low Ground mental and neurobehavioral circuit is also active among all the leading participants.[13]

The Low Ground can also encourage addictive or chronically dysfunctional activity: drinking too much, eating too much, quarreling with someone you're close to, or any other habit that provides immediate reward and relief but makes it harder to reach your true self. One reason for this is the physically pleasant sensations associated with this circuit. The brain's reward center is embedded within the Habit Center; it can literally feel better to think these thoughts. They also feel natural, as though they're the right thing to think.

Indeed, if an external event makes it likely that you will have to change those thoughts, or leave this cycle, it will feel like a threat. Consider what happens, for instance, if you get a report that sales have dropped rapidly. If you are used to having sales grow automatically, and your thoughts about sales management remain on the Low Ground, you will not want to change. You will be tempted to do anything that promises to get you back to that automatic cycle, as rapidly as possible—in other words, you'll be drawn to make a decision based on expedience. You might cancel a project abruptly, cut costs in a destructive way, reprimand an individual, or approve three projects, even though two are unworthy, just in case one succeeds. Or you might decide to go on an unnecessary trip, just to get out of the office momentarily. Make a decision based upon expedience once or twice, and it becomes easier and easier to do it that way the third, fourth, and fifth times.

The organization can reinforce those addictive habits. As people around you mimic your behavior (a phenomenon that is sometimes called "leadership contagion" or the "trickle-down effect" of leadership behavior),[14] you can come to feel this dysfunctional way of life is natural. Low Ground brain circuits often generate feelings of pleasure when others are perceived as similar to yourself. One of many examples we know of comes from a printing and publishing company. Every day, sales employees are sent an image on their screen with a "happy face" or a "sad face," depending on how their monthly revenues are tracking so far that month. This has the desired effect of focusing their attention on the business, but there are also unintended consequences. On the first day of every month, a big sad face appears on everyone's screen, representing the fact that they are, by definition, starting over at zero. In other words, they are reminded of their own gap in performance—a direct provocation of their Reactive Self-Referencing Center. To reduce those bad feelings, every employee now has a built-in unconscious incentive to get new sales any way they can, even if it leads to deleterious consequences later.

The Low Ground leads many companies into similar straits. People do what they have to do to meet their targets, because the alternative is feeling terrible. As a result, they get into trouble—or, at the very least, fall short of the strategic efforts they need to realize their potential. This dynamic appears to be a factor in many well-known stories of business or institutional failure—including the problems that destroyed companies like Enron and Lehman Brothers and seriously hurt companies like BP, Volkswagen, Wells Fargo, and Equifax.[15] When people habitually feel compelled to feel good, it can often lead to doing bad.

Moreover, leaders who do great things—who inspire beneficial social change, develop great new enterprises, foster groundbreaking innovation, or influence others in valuable ways—are known for explicitly eschewing the Low Ground. Reid Hoffman does exactly that, as described in the passages at the beginning of this chapter.

Certainly, the Low Ground has value. You can't ignore it, nor would you want to abandon activity connected to it. There are often times when you want to pay attention to what you want, what others want, the ways to negotiate a deal, and the political realities of your organization. *Does my boss approve of what I'm doing?* is a Low Ground question. So is *Will my direct reports appreciate me for this?* These are useful questions to ask, but they won't lead to the kind of strategic leadership that pushes an organization past its current boundaries. That's why it's good to cultivate the ability to shift to High Ground activity when it's called for.

Leading from the High Ground

Like its Low Ground counterpart, the High Ground brain circuit connects three major centers of the mind and their associated brain regions. The first is the same Warning Center function that links to the Low Ground, associated with the amygdala, the insula, and the orbital frontal cortex. Thus, the High Ground also channels feelings of urgency. Second, instead of the Reactive Self-Referencing Center, the High Ground circuit connects to a function we call the Deliberative Self-Referencing Center. This is associated with the dorsal (upper) medial prefrontal cortex (dmPFC), a brain region above the vmPFC. The Deliberative Self-Referencing Center is activated by consideration of what others are thinking and evaluations of what future actions they might perform. Questions like *What is he thinking? What is she thinking?* and *What will they do next?* tend to be associated with High Ground mental activity. (In chapter 3, we'll refer to this type of activity as "mentalizing" and discuss it in more detail.)

The third function on the High Ground circuit is the Executive Center, associated with the lateral prefrontal cortex. This function is often associated with a future-oriented perspective and is always cognizant of long-term goals and the ability to realize them. When you reflect on your most meaningful aspirations and plan how you might bring those changes to pass, you generate activity in the Executive Center. Leaders often demonstrate their strategic acumen by exhibiting the attributes identified with this function.

One of those attributes is working memory. For a leader, a proficient working memory manifests itself as the ability to deal with complex situations. The more capable your working memory, the more information you can keep accessible at once. This allows you to rapidly and efficiently compare and contrast a multiplicity of alternatives, which empowers you to think about more complex interrelationships and thus solve more difficult problems or conduct more complex cognitive tasks.

A second attribute is cognitive flexibility—the ability to shift perspectives and adapt to new challenges and opportunities as they arise. As we will see, cognitive flexibility is a key attribute of strategic leadership precisely because it enables rapid shifts in perspective. For a leader, a high level of cognitive flexibility shows up in the ability to manage unexpected events with resilience and aplomb, inspiring confidence in colleagues and seeing ways to "turn lemons into lemonade," as the saying goes.

Finally, as we'll see in chapter 3, the Executive Center is the source of inhibition control (the ability to self-regulate, resisting and managing habitual and impulsive behaviors).[16] One of the people who credits this mental function for its role in self-regulation is Columbia University research psychologist Walter Mischel, the creator of the "marshmallow test" experiments, which linked children's ability to self-regulate with success later in life.[17] Inhibition control is one of the core attributes that makes leaders effective; those who don't have it risk losing the respect, allegiance, and collegiality of their peers and subordinates. When you take your attention to the High Ground, it strengthens all of these attributes.

The combination of these three centers promotes rigorous curiosity about the world around you and enables relatively sophisticated insight about people—yourself and others.

You're on the High Ground when you explore the reasons for complex phenomena: what the people involved are thinking, what they are likely to do next, what you are likely to do next, how you might handle a particular situation, and what you should do with all of this insight. The High Ground also seems to be active when you consider other people, not in terms of what they want but in terms of who they are and what they will think and do. *What are my customers thinking? What are my employees thinking? What are our shareholders thinking? What might they plausibly do next? What do they need to do if our entire endeavor is to thrive? What do we need to do to maximize the probability that this will happen?*

There is reason to believe that the High Ground is particularly engaged when you think about yourself or other people in abstract terms, involving different perspectives than just your own. If you ask an amateur athlete "Are you a good soccer player?," it will invoke the Reactive Self-Referencing Center and thus the Low Ground. On the other hand, the question "Does your coach think you're a good soccer player?" engages the Deliberative Self-Referencing Center and is thus a High Ground question.[18] To engage the High Ground this way, you don't have to be entirely accurate in your perceptions of others' thoughts, motives, and future actions. It's enough to energetically and persistently reflect on what they're thinking or what they're likely to do.

As we noted in chapter 1, this is akin to taking a third-person perspective on first-person experience—seeing yourself as others see you. This might seem, at first glance, to be a form of altruism and empathy. But altruism and empathy, while they might show up in High Ground mental activity, are not necessarily the same.

Consider two very self-involved questions: *Am I paid enough? Am I paid fairly?* Neither one involves much empathy or altruism. But only the first is a Low Ground question. Its answer will depend on whether your needs and wants are satisfied. The second is a High Ground question. Its answer depends on your ability to look objectively at your own situation; it calls on you to think about the pay rates in your organization and how they foster (or fail to foster) everyone's success. It's an even stronger High Ground question if the implicit is made explicit: *Would an impartial observer agree that I am paid fairly?* or, even more explicitly, *Would enough of the people whose opinion I value agree that I am getting the reward and recognition commensurate with the contribution I make?*

As it happens, one of the most significant examples (in our view) of High Ground thinking in leadership theory was derived from perceptions about pay, contribution, and hierarchy. The Canadian management theorist Elliott Jaques, who passed away in 2003, spent sixty years researching and articulating a concept he called the "requisite organization": an organization structured for a good fit between the hierarchy and the individual. His research started in the 1940s, when he worked with the Tavistock Institute in London, one of the first psychological institutes to study group behavior in organizations. While conducting fieldwork at a British metalworking company called Glacier, he began confidentially asking workers what they were thinking as they went about their jobs.

The trade union leaders who had invited Jaques to do this were struggling with the perennial problem of pay inequity: Why would a production engineer deserve a higher salary than an account manager? So Jaques asked people what they thought they *should* be making if the company were really fair. (He called this "felt-fair pay.") He also asked what others around them should make and (if they were managers) what positions their subordinates were capable of handling.

This is a High Ground collection of questions. Jaques didn't ask, "What do you want?" He asked, in effect, "What, in the eyes of a dispassionate observer, would your work be worth?" Since the questions had to do with reward, of course many people answered him in a way that engaged their Low Ground. They talked about the money they wanted to make. But at the same time, there must have been a High Ground pattern of activity involved, because to Jaques's surprise, most people agreed on the answers. Jaques found that roughly 95 percent of the people in any given organization agreed on the relative worth of anyone's role and could make roughly congruent assessments of how effectively each individual fit that role. They could even think objectively about their own felt-fair pay.

"One man would smile knowingly about the $80,000 he got," Jaques recalled. "Somebody else with a similar time span would say, 'I don't know what all the fuss is about around here. I'm getting $60,000 and it feels right.' And somebody else would plead with me, 'Doc, you gotta do something for us. We're getting $48,000 and the company doesn't see how unfair it is.'"[19]

The High Ground also seemed to be at play in Jaques's report on the discussions that followed. He asked the people interviewed: Why are some jobs, objectively, worth more than others? A Low Ground answer would have been "The bosses overpay the people they like, and underpay everybody else." But most people knew that was not the whole story; few bosses, even the most Low Ground oriented, disbursed rewards so blatantly, and most were bound by regulations that wouldn't let them play favorites with wages even if they wanted to. Jaques himself was stumped, until one morning three shop stewards burst into his office to tell him they'd figured it out. The critical difference had to do with time. Factory floor operators were paid by the hour, junior officers by the week, managers by the month, and executives by the year.

During the next thirty years, Jaques and a group of fellow researchers conducted thousands of other interviews at organizations around the world, including companies of all sizes, government agencies, and the U.S. and Australian armies. He continued to find close agreement about felt-fair pay. Indeed, there was so much alignment among employee perceptions of what their colleagues deserved that the consensus represented a reliable proxy for human value. Jaques then used that understanding to refine the insight about time span into a fairly universal principle: the value of every job could be measured by the length of time it took to carry out its longest-running assignment. In a large consumer-products company, for instance, the maintenance operator on a factory floor might wrap up all tasks within a twenty-four-hour period, but a purchasing manager might need up to three months to finalize a contract, and a marketing vice president might take two years to plan and implement the introduction of a new soap. The longer the time span, the greater the "felt-fair pay" appropriate to that position.

The concept of felt-fair pay engages the High Ground precisely because it requires people to consider their own (and one another's) wages in a relatively objective fashion, as a disinterested observer would—fair to oneself and to others. Without ever invoking the Wise Advocate specifically, Jaques's inquiry moved people into that domain. The requisite organization concept is still in play at some organizations—a robust, if unconventional, way of tapping the deeper leadership acumen in people throughout an enterprise.

The High Ground also comes into play when you consider long-term goals. While the Low Ground is concerned with subjective value, the High Ground is concerned with genuine worth: whether your ultimate strategic objectives, and the actions (by you and others) needed to achieve them, are important enough to deserve your sustained attention and all the skill and acumen you can bring to bear. Pleasing people and making yourself happy are not enough for the High Ground; it is concerned with achieving something of substance, something that matters. These types of questions appear to be typical of the High Ground: *What are the consequences of this decision? I may want something, but is it relevant to my true self—for example, to the leader I am becoming? What am I being called to do? What needs to happen, and for which groups of people?*

The High Ground tends to apply broad thinking to emotionally sensitive issues, such as discrimination and identity issues. Asking yourself *Do I feel I'm being treated unfairly?* will tend to move you to the Low Ground, no matter whether the answer is. But if you ask *Would a wise person agree that I am being treated unfairly?* and consider the full repertoire of points of view, that will tend to lead you to the High Ground.

Brain research related to what we call the High Ground is still evolving, and its implications are still being worked out. The social neuroscience field has often linked the parts of the brain that we associate with the High Ground with a broader perspective—with thinking about people as actors in a larger system. *What are other people thinking? What are they likely to do?* Jason Mitchell of Harvard University, for example, has found that the dorsal medial prefrontal cortex (our Deliberative Self-Referencing Center) is often engaged when thinking about groups of people in abstract terms, studying them as a sociologist might.[20] However, if you perceive people to be like yourself and identify with them, then when thinking about them, according to Mitchell, you're more likely to use the ventral medial prefrontal cortex (the Reactive Self-Referencing Center, which we associate with the Low Ground).[21]

Adam Waytz of Northwestern and Jamil Zaki of Stanford have also studied brain activity that we would consider correlated with the High Ground.[22] They have found that this circuit is engaged when you concern yourself with understanding what another person is thinking. In another study, Sylvia Morelli and Jamil Zaki point out that when there is thought about the rewards coming to other people, but no thought of the rewards coming to oneself, the dmPFC (Deliberative Self-Referencing Center) region is activated; this region is associated with the High Ground.[23]

Another fascinating finding, by Oscar Bartra of the University of Pennsylvania, suggests that the dmPFC (of our High Ground) becomes active when there is a strong perception of reward or punishment: something especially good or bad is happening or about to happen.[24] This might suggest why leaders who want to motivate organizational change sometimes evoke a "burning platform," as management experts call it. Something terrible is going to happen, they say, unless action is taken right now. If enough people agree to act on these triggers, they may move to the High Ground pattern of mental activity.

We believe that all recognized leaders—all people in positions of authority—use the High Ground to some extent. Even the leader of a street gang has the authoritativeness that comes from occasionally invoking the link between the Warning Center, the Executive Center, and the leader's sense of self. But some leaders are more likely to occupy the High Ground than others. When you deliberately seek a broader perspective on what you are doing, you take your capacity for leadership to a higher level. The guidance available to you from the Wise Advocate becomes more readily accessible, reliable, and fruitful, and is less likely to be overridden or distracted by impulses received from other parts of the brain (for example, by the deceptive brain messages we discuss in chapter 4).

On the other hand, many of the things you might do as a leader—make flamboyant moves, develop a charismatic presence, move rapidly to solve problems or make deals, and become a gregarious host or a benevolent philanthropist—may not really develop your skill at strategic leadership. Instead, they lead you more toward Low Ground thinking and tactical problem-solving. This may indeed foster success, but that success is not as deep or long-lasting as it would be if the High Ground were engaged.

Imagine asking two executives for advice on a new venture or endeavor. The first executive immediately starts talking about what you'll both get out of it. That type of thinking takes place on the Low Ground. *What is the venture worth? What's in it for me?* The second executive considers how other people from other backgrounds would see it and arrives at a decision based on how it would affect the entire enterprise, not just this venture. This conversation doesn't have to be wishy-washy; it can be just as decisive, if you both have the cognitive capacity and perspective to handle this extra complexity. If you do, this conversation takes place on the High Ground.[25]

Self-Directed Neuroplasticity

Your goal in building your leadership competence is not just to move your attention to the High Ground every once in a while, but to develop your own facility with it, to make it more natural, like second nature. To accomplish this, you apply a mental process that will result in self-directed neuroplasticity—the deliberate focusing of attention in a way that reorients the physical circuits of your brain. Over time, that is, the way you direct your attention with your mind will influence the physical structure of your brain.

During the past thirty years, much has been discovered about the way the physical brain is affected by particular patterns of mental activity. For example, when people are shown a frightening picture, a part of the brain called the amygdala is activated in a way that is visible in functional magnetic resonance imaging (fMRI) scans.[26] This activation is physical, automatic, and fairly universal; everyone who is shown those pictures will have the same physical reaction.

However, the relationship between mental activity (in the mind) and neural connections (in the brain) is complex. Although mental activity is often associated with a physical circuit in the brain, those associations can shift when necessary.[27] Neuroscientist Kevin Ochsner of Columbia University conducted a classic experiment that demonstrated this effect.[28] He showed subjects terribly gruesome photographs of automobile accidents. The people who saw the pictures were understandably upset, and their brains reacted. Activity in the amygdala, which is part of the Warning Center, spiked. Then he asked people to imagine that they were emergency medical technicians—in effect, to play-act the role of paramedics. Even though they had no medical or emergency training, taking on this role allowed the subjects to almost completely ablate the amygdala response to these upsetting pictures.

A related phenomenon has been noted among players who suffer head injuries in sports such as ice hockey or soccer. Suppose that two players take a hit from a puck or a ball; one expects it and is prepared for it, but for the other it's a complete surprise. The player who has prepared for the hit—who has framed it as an expected part of the game—is much less likely to suffer a concussion, and the trauma is likely to be less severe.[29]

Further evidence of the complexity of the mind-brain relationship comes from some examples of people who have experienced brain damage. When an injured person receives training intended to refocus his or her attention on tasks designed to improve impaired function, the enhanced capacities developed can relocate to other parts of the brain.[30] Moreover, problems such as addiction, depression, and obsessive-compulsive disorder often elude or resist purely physical ways of addressing them.

All of these examples involve neuroplasticity: the individual makes a choice, whether conscious or guided, to react to the world in a particular way, and that choice affects the physical brain response. In this context, neuroplasticity is one of the most significant aspects of the mind-brain relationship. Indeed, it's one of the most significant aspects of being human. The term "neuroplasticity" refers to the ability of neurons to forge new connections, and the ability of mental activity to rewire the brain this way. This core insight is often expressed as an aphorism credited to Canadian scientist Donald Hebb, who discovered the principle in the 1950s: "Neurons that fire together wire together." In other words, parts of the brain that are continually activated together will physically associate with one another in the future.[31]

Neuroplasticity changes the way people think and behave by making it easier and easier to think and behave in new ways. Significantly, the effect of neuroplasticity is strengthened by repetition. The more frequently a pattern of mental activity occurs in your mind, the more entrenched the associated neural pathway becomes in your brain and the easier it becomes to follow that same pathway in the future—in fact, it can become totally automatic. To use one common analogy, it's as if the pathways in your brain were hiking trails; the more often you occupy them, the more trampled the underbrush becomes and the easier it is to walk there. The pathway literally becomes embedded in the surrounding terrain. And so it is with the brain.

Another analogy that helps explain this aspect of neuroplasticity is the way a powerful search engine works. When you search Google, for example, using a particular term or phrase, the software takes note. It also tracks the results that you click on and records your selection of the items presented to you. The next time you use the Google search engine, it will feature more prominently the terms and results that you chose before, because its design is based on surfacing probabilities: it is likely that its assortment of top hits is closer to what you want, each time, than it was before. You get more of what you've already looked for; the results in your future echo the choices of your past.

If you're accustomed to looking for blame, you'll find your attention is routinely attracted to evidence that someone is at fault. If you're expecting attack, you'll be attuned to indicators of threat. Conversely, if you expect to see opportunities, you'll be more likely to notice them. And if you are predisposed to the idea that people generally rise to a challenge with talent and effectiveness, you'll be relatively likely to see examples in daily life.

Neuroplasticity takes place constantly. Many circuits are formed early in childhood, starting with your first mental and physical activity. These circuits continue to deepen, reinforce, and strengthen throughout your life, forming the basis of many lifelong patterns of mental activity. When your environment changes—for example, when you start going to school—that can change the focus of your attention, creating new patterns of mental activity that strengthen new brain circuits. This is how long-term learning takes place. But this mechanism is also the reason why deceptive thoughts and biases are so pernicious; once they take hold, they become embedded in your brain circuits. The general phenomenon of neuroplasticity explains why any new behavior, such as learning to ride a bicycle, to drink when stressed, or to take charge during a meeting, becomes more automatic and thus easier to perform over time. That's how addiction and obsessive-compulsive disorder, among many other human frailties, gain much of their power.[32]

But deceptive thoughts are not destiny; you can move toward more strategic leadership by deliberately focusing your attention that way. This is the type of focus we call "self-directed neuroplasticity."[33] It is the active use of mind-to-brain influence to change your patterns of thinking and behavior. Directing the mind restructures the brain, especially when you repeat the same patterns of mental activity over and over.[34]

As with any new pattern of mental activity, engaging the High Ground can be difficult at first; the Low Ground is more comfortable, and changes in neural activity can be stressful and exhausting. But as you move more of your attention to the High Ground, neuroplasticity makes it easier and easier to stay with it. The power is in the focus.

If you want to strengthen your strategic leadership capability with self-directed neuroplasticity, taking the High Ground once or twice is not enough—not even once or twice a day. Doing this involves the deliberate use of a principle called "attention density." Attention density is the repeated activity of making choices (whether wise or foolish) about where and how you are focusing your attention. When these choices are sustained in consistent ways, over minutes, hours, days, weeks, months, and years, it literally rewires the brain (for better or for worse). These sustained, consistent changes in patterns of attention, especially when linked to changes in behavioral response, gradually cause profound, empowering changes in patterns of brain activity, and their influence accelerates as the patterns become stronger. When the changes are consistently High Ground oriented, the High Ground circuits are strengthened. With the strengthening of these brain patterns, the choices about focusing attention in mindful, empowering ways become much easier and more automatic.

There is an organizational analogue to this aspect of self-directed neuroplasticity. The organization's culture, embodied in the thinking and conversations that people have, influences the more tangible aspects of the enterprise: its hierarchy, work-flow design, incentives, and information flows. These more tangible aspects, like the physical brain, also deeply influence the direction of activity in the culture. Many people try to change organizations by shifting the "org chart" or making other tangible changes, only to find that the culture resists. It is usually more productive to address the cultural messages first— for example, to talk informally and openly, without reprisal, about the deceptive messages in the organization and why those perceptions, concerns, and feelings are holding people back. Holding that conversation regularly can change people's reactions, which changes the organizational atmosphere and climate, and that allows changes in formal structures like the org chart to take hold more effectively. Just as it is usually more productive to address the patterns of mental activity to change the circuitry of one's brain, it is more productive to address an organization's patterns of conversation to change its formal structure and strategic focus.

As a leader, taking the High Ground can put you in a beneficial cycle of mind-brain interaction. The High Ground patterns of mental activity become visible to others. As we'll discuss in chapter 7, people who report to you continually watch how and where you focus your attention. That's especially true if you are aware that you are doing this deliberately and call others' attention to it. This invokes leadership contagion, and others focus their attention the same way. The result is a change in the culture that accelerates the impact of your own High Ground. For this reason, it's vital that you, as a leader, pay close and mindful attention to your attention: what you expect, whether your expectations are fulfilled or not, and how your expectations influence your choices, attitudes, and actions. You also need to pay close attention to how your choices, attitudes, and actions influence the way others focus their attention as well. In chapters 4–7, we'll discuss a step-by-step process for doing this.

There are signs that complex systems can only be managed from a High Ground perspective and that people genuinely appreciate working for a boss who consistently deals with them from the High Ground. Indeed, business leaders who spend most of their time on the Low Ground are unlikely to break free of the conventional wisdom of their industry. Strategic insights—considerations of the purpose of the enterprise and the long-term value it brings to the world—are more likely to emerge when you consistently occupy the High Ground.

Finally, by moving your mental activity to the High Ground, you become a more active participant in your own improvement, and the improvement of your organization. You are deliberately using the part of your mind that can effectively direct your brain to assess what others are thinking and how they will act under various circumstances and scenarios. And as we will see, there's another big payoff in training your mind this way: it helps move you to a still Higher Ground (as we call it) in which transformational leadership becomes an ingrained way of life, abetted by the internal voice we call the Wise Advocate.

Reflection

When have you seen someone occupy the Low Ground—making a decision based on expedience? How have those decisions typically turned out?

When have you, or someone you observed, moved to the High Ground—making a decision based on a third-person perspective and the kind of guidance you might get from a Wise Advocate? How did that decision turn out?

When making a decision, how do you balance what others want immediately against what is needed most for the situation as a whole?

How do you think about value—more in economic terms or in terms of farthest-ranging consequences of decisions?

What types of decisions are best suited to the Low Ground? What types are best suited to the High Ground? At the moment you're making a decision, how do you know which is best?

CHAPTER THREE

Finding Your Inner Voice

THE EIGHTEENTH-CENTURY ECONOMIC philosopher Adam Smith, best known for his foundational book *The Wealth of Nations*, spent his last two decades considering the problem of virtue in capitalism.[1] The vitality of the industrializing world was based on the good faith of energetic, creative people, acting individually. But no human society had ever resisted the temptations of corruption and exploitation—the propensity of wealthy and powerful individuals to overrun constraints, exploit others, and set up systems that favored them unfairly. How would capitalism survive?

Smith said that the two most obvious means, legal regulations and community censure, were not adequate. Legal regulations tend to be uncomfortably blunt instruments of morality. There is always a way to get around legal regulations, and those who comply often bear unreasonable costs. As for community censure, even the most enlightened community has blind spots, and some fall prey to highly unsavory values. Both regulations and public shame diminish productivity and entrepreneurial vitality. Yet what else could hold the inevitable waves of robber barons in check?

Smith's other famous work, *The Theory of Moral Sentiments*, first published in 1759 and significantly expanded in 1790, proposes a solution principally based on what he calls the "impartial spectator." This is, Smith said, a voice within the mind oriented not just to the individual's desires, needs, and success, but to the overall long-term value of the entire system. It is a kind of narrative voice, a storyteller from within, asking the right questions to help you see yourself and your actions as others might see them. Smith wrote that this voice has the dispassionate perspective of a clear-minded observer and the authoritative voice of a reasonable guide, seeking moderation and tempered responses to life's vicissitudes:

> Should [our] passions be, what they are very apt to be, too vehement, Nature has provided a proper remedy and correction. The real or even the imaginary presence of the impartial spectator, the authority of the man within the breast, is always at hand to overawe them into the proper tone and temper of moderation.[2]

Smith was writing solidly within a tradition of psychological and meditative philosophers who explored the nature of the human mind by examining their own recurring patterns of thought and feeling. He described his impartial spectator as a construct available to anyone who calls on it, an inner voice speaking on behalf of your best interests, helping you move closer to virtue, a voice representing the point of view of a well-informed and genuinely mindful outsider, but part of yourself and directed to yourself.[3] Why should leaders bother with that kind of internalized judgment? Because, Smith said, in the long run it is the key not only to their impact and influence, but also to their happiness, confidence, and self-worth.

As Adam Smith scholar Ryan Hanley points out, the impartial spectator was a key aspect of Smith's attempt to address the inherent ethical limitation in the culture of market economics. Though individual actions, conducted by people seeking their own individual rewards or praise, will often add up to the benefit of all, the invisible hand doesn't always work out. Smith posited the impartial spectator as a counterbalance. It takes its strength from the natural human desire for praiseworthiness—the desire not just to receive praise from others but to deserve it, to be worthy of it, and (if you're a leader) to leave a legacy based on your value to the world at large.[4]

Yet even the impartial spectator, in our experience, is not sufficient for strategic leadership. The impartial spectator, at least as Smith describes it, has one limitation: being dispassionate means being only a spectator, not a true advocate for a better life, a better solution, a better world. To be a truly strategic leader, you need to find a source of genuine strength and character, a source of care and commitment to you, that can stand with you on behalf of your long-term goals. For us, this construct goes beyond what the impartial spectator does, and we use a different name to convey this added level of depth: the Wise Advocate.[5]

We define the Wise Advocate as an aspect of your mind, experienced as an inner presence that you can access, to a greater or lesser extent, whenever you are receptive to it. It can manifest itself (or be called to manifest itself) as a guide or adviser within your mind. It knows what you are thinking, can see deceptive brain messages (and organizational messages) for what they are and where they came from, understands how you feel physically and emotionally, and is aware of how destructive and unhealthy automatic and habitual responses can be—for you, the people around you, and your organization.

You turn to the Wise Advocate to see the bigger picture, including your inherent worth, capabilities, and accomplishments, without flattering yourself or demanding self-esteem or validation—but without criticizing or demeaning yourself. It is not just interested in what is good for you, but what is good *from* you. It is the part of you that is seeking your true self, looking to make decisions on behalf of the long-term best interests of all the systems you care about: your family, your community, your workplace, your enterprise, and more.

Most important, this construct in your mind wants the best for you; it believes in you and cares for you. It encourages you to value your true self and make decisions as a strategic leader, rather than a transactional one. It continually advocates for values and actions that are in your best interest in the longer term, aligned with the High and Higher Ground.

The Wise Advocate is often described through metaphor: a cricket or an angel on your shoulder. But behind the construct is a recurring mental phenomenon. When you repeatedly pay attention to it, you induce self-directed neuroplasticity, and you can rewire the pathways of your brain in ways that significantly enhance your perspective. The most accomplished leaders, from the earliest human history to the present day, have appeared to understand this. By managing their attention to achieve more significant goals, they move their mind more frequently onto the High Ground, and they strengthen their Wise Advocate accordingly.

Although the term "Wise Advocate" was first introduced by this book's author Jeffrey Schwartz (in *You Are Not Your Brain*), the concept has a rich history in philosophy and literature. People have used constructs like this for centuries to transcend difficulties and short-term temptations and to wrestle with issues that at first seem too great to handle. The Wise Advocate can be traced to the ancient Greek concept of *paraclete* (*parákletos*)—an advocate or counselor who stood with an individual to help manage difficulties and distress. The word *paraclete* has been interpreted to mean helper, comforter, advocate, counselor, encourager, strengthener, and friend. The Greeks also associated this construct with the quality of *sophia*, or wisdom; medieval Christian mystics adopted the term *hagia sophia* (holy wisdom). In the Enlightenment, Adam Smith developed his version as the impartial spectator; in pop culture, it was the inspiration for a talking cricket in the original story of Pinocchio, which became Jiminy Cricket in the Disney version.[6]

Philosophers have theorized for centuries about the nature and origins of this inner voice. It has been referred to as a conscience, an inner light, a voice of wisdom, or a moral compass. Christians associate it with the Holy Spirit; Jews, with the "still small voice" of God speaking to Elijah;[7] Buddhists, with the concept of mindful awareness. For some individuals, the Wise Advocate represents the judgment of future generations; for others, it's the experience of being in touch with the universe; for still others, it's simply the awareness that, when looking at the situation from the outside, there is a deeper, broader perspective than simply one's own. This kind of clear-minded, dispassionate, but caring perspective yields the same benefits no matter what you perceive its source to be. Later in this chapter, when we discuss deep mindfulness as an inherent element of the Higher Ground pattern of mental activity, we are explicitly talking about the ability to draw on this source of (among other things) long-term insight and strategic leadership.

While the impartial spectator tends to see virtue in terms of praise-worthiness, the Wise Advocate goes further. It might advise you to be careful with your assumptions about your own praiseworthiness. Your sense of your own progress can go astray, and the higher you rise in an organization, the more likely it is that people around you, trying to please you, will reinforce the deceptive messages that are holding you back. Leaders are particularly prone to receive praise that, if heeded, can lead you back to Low Ground thinking.

Businesspeople often talk about the commitment necessary for success. The Wise Advocate can help a leader realize what that commitment means. It is not just a matter of evoking people's "passion," or "compliance with the strategy," or even their "strategic awareness." It involves understanding what people are thinking, what they are likely to do next, where your own reactions come from, and what has to change—in all of this, including your own reactions—to reach the future you want to create.

One may never fully know where the wisdom of the Wise Advocate comes from, but the experience of it transcends most day-to-day preoccupations. Toward the end of his life, as he continued to write about the impartial spectator, Adam Smith began to write about this. In a passage on the most salutary aspect of the impartial spectator, Adam Smith scholar Ryan Hanley writes:

> Far from mere rhetorical flourish, Smith introduces a concept that does crucial work in his theory of virtue. In suggesting that ethical activity is not limited to the sphere of mere intersubjective human judgments but is rather grounded in a plane that ultimately transcends human judgment, Smith testifies to his belief that the consideration of ethical activity is inseparable from consideration of what lies beyond human action. . . . To put it in terms that have been applied to Aristotle, Smith's recognition of the possibility of "something divine in man" is evidence of his awareness of "the ultimate dependence of morality upon a transmoral good."[8]

You don't have to believe in "something divine" to tap into that voice. But you do have to listen actively for it. This voice in the mind generally doesn't speak on its own. You usually have to call upon it, to seek it out. Indeed, when you first try to connect with this part of your mind, it can seem elusive. Much of the transition to strategic leadership involves learning to be still and receptive enough to invite your Wise Advocate to be present in your mind. Only over time, as you continue to call on it and thus engage the High Ground and the Higher Ground with your mind, does the Wise Advocate become a habitual presence. And by making it a habitual presence, bringing it out again and again, you develop your innate capability to reach deeper and more profound levels of truly transformational leadership. Strategic leadership, in this context, is your ability to make your actions in the outside world correlate with the awareness you gain from your dialogue with the Wise Advocate.

This aspect of mental reflection—actively listening for guidance, rather than assuming it will automatically come to you—is often overlooked. One apt metaphor is the experience of taking up physical exercise, such as running or weight lifting, for the first time. If you're not an athlete, it comes as a surprise that your muscles send you messages. When you squat bearing weights on a barbell, your glutes express a sensation. But the response is much less dramatic than you expect; you have to listen for the signal, and focus your attention on it, to tell which muscles are affected by any given exercise. You have to learn to recognize the difference between an ache in your hamstring and one in your knee, and between a constructive ache, signaling that your muscles are rebuilding themselves, and the kind of pain that means you have injured yourself. If you don't get attuned to these signals, you're at risk; ignoring the signal of an injury can compound it in ways that hobble you for months. However, once you learn to recognize the signals from your muscles, they become a guide to better exercise; you can tell, from those signals, which muscles you've neglected, which would benefit from heavier weight or more repetitions, and which simply need to be stretched. Your own body's signals are a more reliable guide, once you learn to recognize them, than any video or exercise book would be.

The same is true of the Wise Advocate. Once you have learned to consult it for guidance, and to recognize its signals, then you have a valuable source of insight at your disposal. Its suggestions—the realizations that you come to when considering things this way—are the most profound, reliable guides to better leadership and behavior that you can find. But it takes practice and receptivity. In short, the choice to attend to the Wise Advocate regularly—to focus your mind in that way, day after day—is one of the most important habitual decisions that an accomplished leader can make.

A Deeper Context for the Inner Voice

What, then, are the ways in which you can build this capacity in yourself? That's where the four steps of strategic leadership come in: relabeling, reframing, refocusing, and revaluing. Most serious leaders are on a progression toward greater capability, using some path similar to these four steps. The steps help develop the High Ground patterns of mental activity that we discussed in chapter 2.

In that chapter, we focused on the two most relevant patterns for most leaders: the Low Ground and the High Ground. In this chapter, we'd like to broaden the backstory and describe four patterns of leadership thinking we have observed: Immature Leadership (Lower Ground, driven by instinct and emotion); Transactional Leadership (Low Ground, driven by expedience); Strategic (High Ground, driven by the need to improve long-term results); and Transformational (Higher Ground, driven by ongoing aspiration to make the most of this enterprise). This will help you better understand what is going on in your mind as you move along the High Ground toward an even more mindful Higher Ground.

Table 3.1 spells out the differences among these four broad patterns of mental activity. It also shows how the Wise Advocate is available in each, the types of environmental cues most likely to capture your attention, and the mental activities characteristic of each. In the rest of this chapter, we'll look more closely at the mental activities related to the High and Higher Ground: mentalizing, executive function, applied mindfulness (which combines meta-cognition, or thinking about your thinking, and meta-attention, or paying attention to how you pay attention), and deeper forms of mindfulness, which we refer to as meta-awareness. Each of these is a skill you are building within yourself; as your prowess grows, the Wise Advocate in your mind becomes more prominent and valuable as a source of insight, and your leadership capacity may grow accordingly.

You can regard this as four successive floors of a building; the higher you rise, the greater is your legitimacy as a strategic leader. Even on

TABLE 3.1
Four Patterns of Mental Activity for Leaders

Pattern	Leadership	The Wise Advocate	Salience (What is Most Likely to Capture Your Attention)	Activities that Invoke this Pattern	Mental Narrative
Higher Ground	**Transformational leadership:** Making the most of your organization's potential and helping it become what it is called to be.	You consciously and habitually consult your Wise Advocate and play the role of the Wise Advocate for the larger organization, all in the process of developing your true self.	What is relevant to your most significant goals, and those of your organization; the development of your true self.	**Deep mindfulness:** Meta-awareness	Am I consulting my Wise Advocate? How would the loving guide in my mind regard what I am thinking and doing? What does it suggest? Thinking about what you're thinking; Paying attention to what you're paying attention to. What am I thinking? What am I paying attention to?
				Meta-cognition and Meta-attention	
				Applied mindfulness: *(associated with High and Higher Ground)* Consulting the Impartial Spectator Mentalizing about yourself	How would an objective, highly credible observer regard what I am thinking and doing? What am I thinking? What am I likely to do next?
High Ground	**Strategic leadership:** Addressing problems that are too complex for an organization to manage easily; helping the organization transcend its limits; being able to rapidly move between High and Low Ground.)	You tend to consult your Wise Advocate when you make important decisions and manage others.	Patterns of behavior and systems, including those beneath the surface; the thinking of people around you.		

		Executive function:	What does this look like if viewed from a different perspective? How does this relate to what I already know? Should I refrain from acting on this impulse?
		Cognitive flexibility Working memory Inhibitory control (Free won't)	
		Mentalizing	What is he or she thinking? What are they thinking? What are they going to do?
Low Ground	**Transactional leadership:** Making deals and solving problems; getting what you want and giving others what they want; making expedient decisions.		
	You may consult your Wise Advocate when it seems relevant.		
	The problems that need solving right now; the urgent issues that you feel you can't ignore.	Subjective valuation	What do they want? What do I want? Is it valuable to them? Is it valuable to me? Is it relevant to me?
		Responding to habit	This is the way we do things around here.
Lower Ground (instinct and emotion)	**Immature leadership:** Driven by sensations, cravings, and the need for immediate gratification.		
	You tend to ignore your Wise Advocate.		
	Your impulses, the most obvious external events, and the threats and rewards you perceive.	Following impulses and emotions	Something is wrong! I'm attracted to this! I'm happy, sad, surprised, angry, disgusted, desirous, or afraid! I must say or do something about it!

the lowest floors, the Wise Advocate is accessible to you and can help you move to a higher level. Consulting your Wise Advocate is a vehicle for upward mobility.

- On the Lower Ground, the realm of instinct and emotion, you call on the Wise Advocate out of desperation, when you feel you've hit rock bottom. It can help you rise to more effective mental activity.
- On the Low Ground, you call on the Wise Advocate when you feel moved to do so, and you have to force yourself to do it. But it opens you to a broader perspective.
- On the High Ground, you call on the Wise Advocate in a self-aware way, knowing that this will lead to your improvement and advancement.
- On the Higher Ground, you call on the Wise Advocate as a matter of course. It's what you naturally do.

In any of these levels, you never lose your access to your more basic instincts and emotions. You can summon them at the most rarefied levels of Higher Ground, and you may use them in more constructive ways—for example, to express measured indignation or to take pleasure in someone's company—without becoming trapped by the mechanisms of the brain. And let's be candid, sometimes it is necessary to express emotion or to follow instincts to compete successfully in the short term. Street fights do arise in corporate life.

Let's look briefly at the lowest floor: immature leadership, driven by instinct and emotion. In this pattern of mental activity, compulsion and the gut brain (animal instincts) are closely linked to subjective valuation. It appears to be a habitual pattern among leaders who fall into self-aggrandizement, envy, and grandiose ambition. The associated brain circuit involves the ventral striatum, which contains the nucleus accumbens, the center of pleasure signals that we briefly mentioned in chapter 2. Thoughts about what you want and need tend to degrade, replaced by strong feelings of craving and disgust, operating on a primitive mammalian level. *What am I attracted to? What am I repelled by?*

Much of the field of neuromarketing has focused on this brain circuit and related patterns of mental activity. The brain areas are physically located below those of the Low Ground when the head is held upright, so we sometimes think of this as going "over a cliff." Many instinctive attitudes are given free rein when you fall over the cliff; you are more likely to dehumanize your opponents, to think about others as illegitimate or unnecessary, and to think it's a virtue to exploit or harm others. You will tend to expect legitimacy as a natural outcome of your position, and you'll receive it so long as people perceive you are operating on behalf of their interests, or if they are afraid to confront you.

In our experience, few experienced organizational leaders spend much time in this stage, and those who do have typically slipped here under the influence of some form of addiction. Of course, there are always some executives who rant openly, mistreat others when they feel provoked, and treat the organizations around them as vehicles for their own aggrandizement. This type of narcissist behavior, however, is less widely accepted than it used to be, at least in business. If an organization is transparent in its operations, if employees participate in making decisions, and if people are recruited and promoted on the basis of their professionalism (which is the case in most enterprises today), then immature leadership will not play well. Most businesspeople have learned to manage their primal impulses and cravings, at least enough to cross the threshold into professionalism. Otherwise, most of them could not advance to any significant leadership role.

As we saw in chapter 2, however, most businesspeople are still prone to the transactional leadership of the Low Ground. They still make deals designed to produce expedient results: giving people what they want in order to achieve near-term goals and rewards. If you want to move beyond this as a leader, however, then several habits of mind will help you move to the High Ground. The first is the ability to see others more clearly—an activity known to neuroscientists as mentalizing.

Mentalizing: Sensing What Others Think

Social neuroscientists have studied the activity called mentalizing in some detail. It always involves thinking in depth about other people in a particular way: reflecting about what they are thinking and what they are likely to do next. The research we have found on mentalizing demonstrates clear links to the brain functions associated with what we call the High Ground. There are several reasons to conclude that the activity of mentalizing is one of the most common elements in High Ground thinking and a good way to move your leadership style in that direction.

Mentalizing is thinking in depth about other people and making inductions about them. When you mentalize, you are reflecting on people's essential nature and what will happen in the future. You consider people dispassionately, trying to figure them out as if they were characters in a novel or a film. *What is he thinking? What is he likely to do next? What is she thinking? What is she likely to do next? What do they mean by this? What makes them tick?*

Another term for mentalizing is "theory of mind." In psychology, this means having a view about what another person thinks and how he or she will behave in particular situations.[9] To come to an accurate theory of someone's mind—to mentalize—involves reflecting on the person's motivations, and on what she or he is thinking. In a highly social environment, mentalizing is a survival skill, and those who don't like to do it (such as high-functioning autistic people) tend to suffer accordingly. Some people have an easier time with it than others; their more nuanced understanding of other people helps them occupy leadership positions.

In a typical study of mentalizing, psychology researchers ask subjects to look at groups of pictures illustrating simple stories or to read passages describing simple situations. Then they are asked to explain the behaviors in the pictures or stories. For example, a subject may be shown a photograph of another person's face and body posture, given some contextual information, and asked what that other person was thinking about when the picture was taken, or what that person was intending to do. Somebody who answers the question accurately is perceived as being adept at mentalizing.

When these experiments add an fMRI component, mentalizing consistently activates the Deliberative Self-Referencing Center (dmPFC), which is part of the High Ground circuit.[10] This is one of the most established neuroimaging findings for the aspect of the medial prefrontal cortex we call the Deliberative Self-Referencing Center. For example, Jason Mitchell's studies of these areas of the brain have found that mentalizing engages them, whether the subject is thinking about an individual or a group.[11] Participants know that groups are not individuals, but even so, when they attribute intention to a group of people ("Our employees are looking for a straight answer about the company's strategy"), it engages the parts of the brain associated with the High Ground.

A good example of mentalizing in business comes from Mark Bertolini, the CEO of Aetna, a major U.S. health-insurance company based in Hartford, Connecticut. Bertolini has long been lauded as a strategic leader; for example, management consultant Ram Charan singled him out as prescient when he helped engineer a merger in 2017 between Aetna and the CVS pharmacy retail chain, a game-changing move for both companies.[12] But that wasn't his most prominent mentalizing move.

Bertolini made headlines in early 2015 as the first chief executive of a major company to offer his lowest-paid employees an across-the-board wage increase, to sixteen dollars an hour. For the 5,700 employees who stood to benefit, this meant an average pay increase of 11 percent; some saw an increase of 33 percent. Most were call-center employees; many were single parents who had children on Medicaid and food stamps even while they worked a full-time job. This was arguably the most visible wage hike by a chief executive since 1914, when Henry Ford doubled his assembly-line workers' pay to five dollars a day. And it had a similar rationale: by raising minimum wages, Bertolini was setting an example that would lead, he hoped, to more general prosperity, so that more people would be able to become his company's customers.

Now listen to the thought process underlying his decision (as voiced in a published interview).[13] First, he spent a great deal of time and effort trying to learn more about what his employees were thinking:

> I became active on social media. We have an internal network called Aetna Connect, and I'm constantly talking to the employees on it. They also talk to each other. More and more often, I saw people online saying, "I can't afford my benefits. My healthcare coverage is too expensive." I heard the same thing in site visits. When I visit an Aetna office, after the town meeting where I speak, I try to go to every cubicle in the building and shake everybody's hand. I ask them what they're up to and how they feel about it here. The same message came through.

Next, he looked closely at the other people in the organization, particularly those in management levels or human resources. What were they likely to think, and what were they likely to do?

> [The leaders of Aetna business units were] starting to agitate about the income inequality problem. Some of them were worried about turnover, and being able to keep people motivated on the front line. After we had looked at a number of options to help our lowest-paid employees, I finally said, "How about we just pay them more?"
>
> [But to do that,] we needed more data. I asked the HR team to build a profile: How much did the lowest-paid people at Aetna make? What did their healthcare coverage look like? What were their out-of-pocket costs? How hard was it for them to get by?
>
> It took months to get this information. The organization really wasn't ready to talk about it, but I kept pushing. . . . I asked for total turnover costs. How many people leave involuntarily? How much does it cost to hire their replacements? How long does it take to train the new recruits? We looked at absenteeism, rework, productivity, dissatisfied employees, and our net promoter scores [a measure of survey respondents' enthusiasm] in recruiting new employees.

He and his colleagues thought carefully about the form of the raise. The point wasn't just to give them more money, but to address the fundamental issues keeping those employees from making a commitment to the company. This meant not just asking employees what they wanted, but thinking intensively about what they were thinking and how they were expressing it. He made the connection, for instance, to health-care coverage; in addition to raising salaries, he lowered premiums on high-quality health insurance for them. ("Our objective was to raise their personal disposable income as high as we could without breaking the bank.") And he grouped them, in his mind, with Aetna's customers:

> [Our customers and employees] really want safety. They're not saying, "I can't wait to get insurance so I can run off to the doctor and spend somebody else's money." They're solid working people trying to take care of their families. I want us to really care about those people.

When I was growing up, my dad worked as a pattern maker in the auto industry. Since the models changed every five years, he only worked six months every year. My mother was a nurse in a pediatrician's office. That's how we got our healthcare paid for. Our market is families like that.

Even the way Bertolini recalls making the announcement shows him thinking about what other people are thinking:

I went down to our largest call service center in Jacksonville, Florida to announce it. . . . Everybody was wondering why I was there. "He's retiring." "The company's been sold."

You can think of this as ordinary good management; every senior executive should pay attention to what employees are thinking. But in our experience, it's very rare—not just for a senior executive, but for anyone—to spend this much attention and trouble figuring out what other people are thinking and how they might respond, even on important issues. This habit also partly explains Bertolini's reputation as an uncommonly idiosyncratic CEO, especially for the health-insurance industry. He has harshly criticized both Democratic and Republican health-care plans in the United States and outspokenly supported wellness-oriented preventative care.[14]

Mentalizing is not like thinking about people in transactional terms, the way you would on the Low Ground when you ask yourself, *What does the other person want?* True mentalizing, on the High Ground, can actually lead you to dismiss others' concerns after considering them, as Bertolini did for one group of Aetna employees. As he recalled in the interview:

"Some employees said, 'Wait a minute. I worked six years to get to $16 per hour and here you are handing it to them. What are you going to do for me?' It was disappointing to hear this."

Bertolini pointed out that, while he thought about this group and understood how they felt, he wasn't just out to please people or give them what they wanted. His purpose was to raise the floor, to correct a long-standing inequity; people at the next-higher salary level didn't have the same survival issues. A Low Ground approach, driven by subjective valuation, would have given raises across the board—or decided that it would be too difficult to manage the backlash and dropped the idea. Bertolini wanted to do what was right for the company. His purpose was to raise the floor, to correct an ingrained long-standing inequity. (He did address other employee concerns by reframing Aetna's health-care practices to offer enhanced benefits based on household income and wellness and lifestyle practices.)

He also understood that most of the people in the company would take genuine pride in being among the first to correct this problem:

> We had given our top 300 managers a heads-up the day before. We got them on the phone. They said things like, 'This is the proudest moment I've had in 42 years at the company. [Then, after making the announcement,] the place exploded. I had known people would be happy, but I wasn't ready for the raw emotion.[15]

Since then, a number of other companies have followed Aetna's example.

Mentalizing doesn't have to be accurate to trigger the High Ground. Your mental simulation of someone else's mentality—*What are they thinking?*—could be wrong. But if you are genuinely interested, then mentalizing could also involve testing your assumptions, asking people what they are thinking or what they are likely to do next, instead of relying on your guesswork about them. This is the basis of a great deal of organizational learning practice, the sort developed by theorists Chris Argyris and Donald Schon and his colleagues and popularized by Peter Senge in *The Fifth Discipline.* For example, Argyris developed a series of practices around more effective inquiry, in which managers would avoid questions like "What's your proof?" or "What do you mean?" and instead would ask, "Can you help me understand your thinking here?" or "Are we starting from two very different set of assumptions here? Where do they come from?"[16] Questions like those are only meaningful if you already have a theory about what the other person is thinking and if you are willing to test that theory. You don't have to be accurate in your mentalizing to trigger the High Ground; you just have to be willing to think this way.

Mentalizing is also distinguishable from empathy. You don't necessarily have to care about people or identify with them in order to sense what they think in this way. Rather than pure empathy, a good mentalizer tends to have the capacity for what Stanford researcher Jamil Zaki calls "empathic accuracy."[17] This is the ability not only to recognize people's thoughts but to observe and recognize their emotions—their feelings of pain, happiness, frustration, anger, or joy. If you have this kind of radar for other people's emotions, then you can more easily discern both what other people think and what values are driving them.

Mentalizing doesn't involve feeling good (or bad) when others are rewarded, but Sylvia Morelli (director of the University of Illinois's Empathy and Social Connection Lab) and Zaki have linked mentalizing to a form of thinking that psychologists call "vicarious reward."[18] This is a kind of curiosity that is triggered by the experience of seeing someone else benefit. It involves, as they say, "an observer's [inference] about whether outcomes are in fact valuable to [other people]." Rather than comparing their rewards to your own, or being happy for them, you are intrigued by their potential response. *What will they think? What will they do now?*

The Mentalizer's Paradox

There is reason to believe that mentalizing is good for you. It makes you a more effective leader. It connects you to the High Ground. It gives you more insight into what others will do and what they're thinking. And there may be evidence that people accustomed to mentalizing are better communicators than other people.[19]

Emily Falk of the University of Pennsylvania refers to this as the "salesperson effect." In the research she conducted, people were asked to communicate new ideas to others while brain activity was being observed. The ideas that connected successfully tended to come from people with a higher response in areas associated with the High Ground brain circuit. People who were good salespeople were also linked to this circuit. Falk and her colleagues attribute this communication effectiveness, at least in part, to mentalizing.[20]

With all this benefit from mentalizing, you'd naturally expect leaders to pursue it as a path toward leadership—but they don't. The paradox of mentalizing lies in the historical relationship between mentalizing and privilege: because it's seen as something that low-status people do, as people rise in the hierarchy they are less inclined to mentalize; yet executives who cut back on their thinking about other people can lose their edge as strategic leaders.

The research related to this paradox is limited, involving only adolescents and college students, but it resonates with our experience in the business world. The people who mentalize most frequently—those who are, as one study puts it, "more likely to engage in social cognitive processes that aid in understanding how others think, feel and behave"—may be those who see themselves as individuals with low social status.[21] In this study, being "low-status" meant coming from a poorer background, and thus have less spending money and fewer economic options, than others on campus.[22] But the mentalizer's paradox can apply to any form of status inequality; as Robert Fuller points out in his book *Somebodies and Nobodies*, there are as many ways of being labeled low-status as there are human groups.[23]

People who see themselves as lower in status tend to have at least one thing in common: they consistently mentalize about higher-status individuals. For example, one major role that practically requires mentalizing is that of caregiver: watching over elderly or cognitively impaired persons who can't manage themselves. A caregiver has to continually ask, about the person being taken care of, *What is he thinking? What is she feeling? What are they likely to do next?* The same is true of any role that involves serving others (such as assistants, restaurant staff, and store clerks). People who feel they have been wronged tend to think quite a bit about the people who have wronged them: *What they are thinking? What they are likely to do next?* And people who report to a boss are typically involved in mentalizing about that individual: *What is she thinking? What's in his plan? What are they likely to do next?*

From this perspective, you are underprivileged if you feel forced to mentalize—to pay attention to what others are thinking and what they are likely to do next, while they feel no obligation to pay attention to you. When a job or role is considered demeaning, it's often because the people in that position are expected to mentalize more about their customers, their investors, their bosses, and everyone else, while no one pays attention to them in the same way.

The underprivileged don't just mentalize frequently; they are also good at it, often because they have to be. Circumstances make them, as the same research study put it, "more accurate at inferring the emotional states of others, relative to their higher-status counterparts."[24] Thus, employees know not only the immediate preferences of their bosses but also what their bosses care about fundamentally, even when the bosses don't fully know themselves. In some marriages, there is a less privileged spouse who is a consummate connoisseur of the emotional state of the other. Job applicants consider what recruiters care about; salespeople know what their best clients are thinking; the bullied anticipate what their tormentors are likely to do next.

This would seem to put the mentalizers at a disadvantage. But mentalizing is an act that triggers the High Ground and is thus foundational for leadership. If more executives could get past their distaste for mentalizing, they would find it is very beneficial for them as leaders.

Consider the truly strategic leaders you have known. Like Bertolini, they are known for their ability to mentalize—to think about what other people are thinking, and what they are likely to do next, with some degree of accuracy. Because of the intensive, high-voltage way they pay attention in conversation, they tend to give the impression of genuinely caring about what other people think. This is an essential part of what Robert Greenleaf, in his well-known 1970 essay *The Servant as Leader*, calls "servant leadership." He uses the word "listening" to describe the attentiveness a leader gives to what other people are thinking:

> Only a true natural servant automatically responds to any problem by listening first. When one is a leader, this disposition causes one to be seen as servant first. . . . I have seen enough remarkable transformations in people who have been trained to listen to have some confidence in this approach. It is because true listening builds strength in other people.[25]

Many leaders who have this ability acquired and refined it by moving slowly up the ranks to their current positions of authority. Along the way, they often mentalized about people they worked with. These leaders internalized the understanding that they are not the center of the universe; that others can see the same things from very different perspectives. By contrast, people who moved rapidly to the top, or who grew up privileged, can be particularly challenged when they move into positions requiring strategic leadership, because they lack practice in this skill.

People in low-status roles can easily come to resent mentalizing. If you're in that kind of role, you might see it as something you are forced to do, and you may be unaware of the way it has helped you gain awareness. You may never really feel comfortable devoting your attention to what others are thinking and feeling. You might scorn mentalizing as something assistants or subservient people do; you could see it as paying your dues. Moreover, when you better understand other people's thoughts about you, your self-esteem could take a hit. You might also find it innately difficult and stressful.[26] It takes real mental strength, even pain tolerance, and a lot of hard work to be a good mentalizer. For all these reasons, as soon as you get to a reasonable level of achievement and status yourself, you might stop. It often feels like a triumph or a relief to leave mentalizing behind and to have others pay attention to you for a change.

But the hard work and pain of mentalizing is even more beneficial when you feel you no longer need to do it. After you rise to higher levels of responsibility and authority, people tend to shield you from criticism. You no longer receive signals about what others are thinking and feeling, and you lose touch with that aspect of your environment. You no longer feel you need what psychologists Susan T. Fiske and Eric Dépret call "diagnostic information about others," and you're more prone to make false assumptions and stereotypes about other people.[27] Your internal Wise Advocate no longer has ready access to that information. You begin to think that maybe you *are* the center of the universe, since others treat you that way. And ultimately you may regress from the High Ground back to the Low Ground.

Although this phenomenon has not been studied clinically—to our knowledge, there has been no longitudinal study of mentalizing by individuals over the course of a lifetime—we have seen it in people we know who rise in an organization. They shift their own mental activity in a way that makes them less knowledgeable about the people in their environment. They reduce their own long-term capacity for becoming highly effective leaders. Their Low Ground gets stronger, and their High Ground is less pronounced. We think this paradoxical aspect of mentalizing may be linked to the growing number of reports about the debilitating effect that gaining power has on leaders' intellectual capacity.[28]

If the mentalizer's paradox leads to debilitated capacity, then the opposite should be true: leaders who continue to mentalize should retain their decision-making vigor. We haven't seen any substantiation of this, but we have met a number of senior leaders in business who seem to have kept mentalizing even as they gained status.

A few extremely successful politicians (Ronald Reagan and Bill Clinton come to mind) have been recognized for their well-practiced ability to understand what others are thinking and what they are likely to do next. They give the impression of understanding what other people think because of the intensive, high-voltage way they pay attention in conversation.

You can also see the power of mentalizing in some stories of successful business entrepreneurs. Moses Znaimer, the founder of the CityTV network (a Canadian predecessor to MTV, with a monopoly broadcast system), the Bravo channel, and ZoomerMedia, started his career as a reporter doing person-in-the-street interviews, asking people directly what they thought about current events and what they intended to do about them.[29] IKEA founder Ingvar Kampraad was known for his keen interest in his middle-class customers and for anticipating what they were thinking but couldn't quite articulate. He would prowl through IKEA stores asking people, "How did we disappoint you today?"[30]

When leaders practice mentalizing all through their careers, like Natalie in chapter 1, their High Ground gets stronger. Mentalizing, which others view as an obligation and a chore, becomes an opportunity for them to develop and grow as leaders. Organizational researchers Tom Malone (at MIT's Sloan School) and Anita Williams Wooley (at Carnegie Mellon) studied team intelligence (as indicated by group problem-solving tests) in relation to a variety of other factors. The most consistently strong correlation they found was with "social perceptiveness"—the ability to correctly read what other people are thinking and feeling. To evaluate this, they used a test called "reading the mind in the eyes," devised by British autism researcher Simon Baron-Cohen, which Baron-Cohen describes as an advanced test of theory of mind.[31] Charles Duhigg, in a 2016 study conducted for Google, also found that social perceptiveness makes teams more effective.[32] Interestingly, according to Malone, while groups vary dramatically in social perceptiveness, there is no difference between face-to-face groups and groups that meet only online.[33]

All of these reports suggest that mentalizing is a critical factor in the development of strategic leadership. If you can practice it in the spirit of connection to your fellow human beings—and to the patterns of behavior that they move in—you will find it becomes a powerful practice that leads to greater empathy and awareness.

One final twist to the mentalizing paradox: the practice of mentalizing doesn't automatically lead to strategic leadership. Despite all the value it offers, there are many people who mentalize and never rise to more senior positions and never become more influential. Some of them are simply unambitious, but many of them seek influence and never find it. Mentalizing is not sufficiently powerful without its senior partners: executive function and applied mindfulness.

Executive Function

In J. R. R. Tolkien's renowned fantasy trilogy *Lord of the Rings*, one of the most telling examples of failed leadership occurs at a climactic moment. The two most powerful members of the noble order of wizards, Saruman and Gandalf, find themselves on opposite sides of a great war, parleying after a fierce battle. Saruman speaks from the height of his tower castle, where he is under siege; Gandalf and his ally Theoden, the king of the cavalry-like Riders of Rohan (also known as the house of Eorl), sit on horses at the tower gates.

At first, Saruman is reasonable; he speaks like a "kindly heart aggrieved by injuries undeserved." His goal is to persuade Theoden to shift sides and support him, and it seems as though he will succeed. But then he is provoked. Theoden rebukes him, calling him a liar and a corrupter of men's hearts, and Saruman's measured presence falls away:

> To some suddenly it seemed that they saw a snake coiling itself to strike. "Gibbets and crows!" [Saruman] hissed, and they shuddered at the hideous change. "Dotard! What is the house of Eorl but a thatched barn where brigands drink in the reek, and their brats roll on the floor among the dogs?"[34]

At this moment, it becomes clear to everyone present that Saruman is not qualified to rule. A psychologist transported to Middle Earth to observe this scene might say that the wizard lost control of his executive function. Transported by his rage, Saruman could no longer manifest the type of reasoning, problem-solving, and goal-directed planning that we associate with strategic leadership. In this passage, Tolkien taps into one of the great recurring themes in literature: loss of emotional equanimity as a signal of a leader's lack of ability to occupy the High Ground.

In real life, the moments of truth may not be as dramatic, but the dynamic still holds. Those who cannot exhibit high levels of executive function cannot remain on the High Ground very long. Conversely, managing the executive function is one of the key activities that invokes and reinforces their presence on the High Ground. Those who break under pressure, like Saruman, become discredited. At that point, they have a choice: to reclaim their executive function by demonstrating the necessary habits, or to sink further back toward instinct and emotion.

In chapter 2, when discussing the Executive Center, we named the generally accepted three main attributes of this function: working memory (holding complex ideas in mind), inhibitory control (regulating impulses and emotions), and cognitive flexibility (adapting to change and looking at the world from more than one perspective). One indication of cognitive flexibility is the ability to move rapidly and willfully between High and Low Ground—between short-term and long-term priorities.

Another test of your executive function is how you operate under stress—particularly the stress of a situation you have not prepared for. The middle of a stressful situation is not the time to develop your executive function. You have to build it during the easy times, day after day, so that you are prepared to rely on it during the difficult times. Most professionals have learned this discipline in part; they manage themselves, advance through their careers, and build emotional resilience over time. But there are always ways in which even mature individuals fall short; there is always more to do. As University of British Columbia cognitive neuroscientist Adele Diamond puts it:

> Using [the executive function] is effortful; it is easier to continue doing what you have been doing than to change, it is easier to give into temptation than to resist it, and it is easier to go on "automatic pilot" than to consider what to do next. . . . [But these] are skills essential for mental and physical health; success in school and in life; and cognitive, social, and psychological development."[35]

In a comprehensive summary of the research on executive function, published in 2013, Diamond articulates its elements and how they fit together. Working memory is the name given to the human ability to hold information in mind and mentally work with it, even when you can't see or hear it directly—for example, doing math in your head, updating or changing a plan, or synthesizing diverse facts into a common understanding. A robust and active working memory, she says, is "critical for making sense of anything that unfolds over time."[36]

Working memory also supports the other elements of executive function. For example, the ability to inhibit distractions and remain goal-oriented requires input from working memory. In turn, the inhibition of distractions is essential for inhibitory control. This type of mental activity involves not only holding back intrusive thoughts and memories, but also focusing attention and controlling impulsive behavior. (The inhibitory control of attention is sometimes called executive attention.) Together, working memory and inhibitory control enable cognitive flexibility. This is the ability to see a situation from multiple points of view and easily consider new perspectives. It is characteristic of creative people and people who are comfortable with ambiguity and change: people who can rapidly switch from one approach to another and adapt to new challenges and opportunities as they arise. Cognitive flexibility, in turn, has been shown to be a critical enabler of mentalizing (theory of mind), along with still more complex aspects of executive function: reasoning, problem solving, and planning.

It follows that the practice of executive function will strengthen your connection to the High Ground, and ultimately to more strategic leadership presence. But can your executive function actually be improved? Or is your capacity for, say, inhibitory control essentially innate, determined by genetics and very difficult to change?

According to Adele Diamond, the research suggests:

> [It is] trainable and can be improved at any age—probably by many different approaches. Repeated practice is key; exercising and challenging executive functions improves them and thus is beneficial for our mental health, much as physical exercise improves our physical fitness and is beneficial for our bodily health.[37]

Practice does help a great deal, as we'll see in our four-step model in chapters 4–7. In addition, your attitude about this question has an effect. Psychologist Carol Dweck's research suggests that a "fixed mind-set"—in which you believe you and others are constrained in your abilities and will never overcome those constraints—leads to a draining away of executive function. Conversely, a growth mind-set—in which you believe there is always the opportunity to improve—helps expand your cognitive flexibility and other aspects of the executive function as well.[38] This is a beneficial, mutually reinforcing form of what we'll describe in chapter 4 as "meta-cognition"—thinking about your thinking. If you believe that you're capable of continually improving your executive function, your executive function's cognitive flexibility is more likely to be active and to strengthen that belief.

It's fortunate that executive function can be improved, because it is a key component of High Ground mental activity. Indeed, one of the beneficial aspects of the four steps of strategic leadership—relabeling, reframing, refocusing, and revaluing—is that they emphasize and help develop executive function, and emotional regulation in particular. James Gross, one of the most prominent researchers in the field of emotional regulation, was also a pioneer in codifying this type of inner work, which he called "reappraisal."[39] Our second step, reframing, is an applied version of reappraisal.

Executive function is necessary to maintain equanimity and avoid visible anxiety, or you cannot play the role of Wise Advocate in the organization around you. Nor will you have the self-discipline to rise above expedience. But executive function, like mentalizing, is not sufficient in itself. It's not enough to be the master of your externally directed thoughts and emotional responses. You also have to broaden the way you think—to think about your own thinking and become aware of the causes and effects of your thoughts. This is known as meta-cognition and meta-attention; we also refer to it as applied mindfulness.

Applied Mindfulness

Applied mindfulness is the form of mindfulness that most people know: the pragmatic application of contemplative practice. Its goals are self-knowledge and self-regulation, through thinking about your thinking (meta-cognition) and observing the way you pay attention (meta-attention). For the High Ground, applied mindfulness generally involves consulting the Wise Advocate and mentalizing about yourself. It can also be a gateway to the deeper mindfulness and meta-awareness of the Higher Ground.

Millions of people have been exposed to some form of applied mindfulness in the context of meditation. You sit in a comfortable but upright position, your spine straight, perhaps with your legs folded. You draw your attention to some regular aspect of your experience—in one common and extremely beneficial form, you focus your attention on your breathing. Each time your mind wanders off on a tangent, you catch yourself and bring your attention back to your breath. As you do this regularly, you develop new mental skills. For example, you gain an enhanced awareness of thoughts moving through your mind. This practice also induces self-directed neuroplasticity: it changes your brain.

The state of mind that you enter is an enhancement of ordinary mental life. The effects of this enhanced mental state actually change the prevailing electromagnetic waves emitted by the brain.[40] Even limited practice increases your ability to focus attention and significantly coordinates and harmonizes connectivity across different parts of the brain.[41] With more training and practice, the proportion of faster gamma waves emitted by the brain increases.[42] These increases in gamma wave power and amplitude have been correlated with enhanced creativity and intuitive thinking.[43] In this state of mind, you gain an increasingly clear awareness of the thoughts moving through your mind; you can watch them moving, almost as if they were being thought by someone else. This sort of perspective is key to the impartial spectator and especially to the Wise Advocate.

Training and guidance in meditation can deepen and stabilize establishment of the practice. Most of the training has to do with ancillary factors: posture, attitude, increased awareness of cues that draw you back to the object of meditation. But the real key is regular, repeated practice of even a few minutes a day. For many people, sitting in a room with other people in contemplation makes the experience more accessible, or even more profound. For others, it is meaningful to be triggered by a recorded message—a sound environment—or by a biofeedback device. The most consistent support is a timer: in contemplative practice, it's important not to be distracted by wondering whether you have meditated long enough. All of these facilitating aids may be helpful, but what matters most is the consistency of practice.

One of the first things you learn in this practice is that, no matter how experienced a meditator you are, your mind will wander, at least some of the time. Mind wandering is a passive and unavoidable part of the process. This is because your brain is typically in control, being attracted to whatever is most habitually resonant. Inevitably, the internal monologue of your brain—those pressures of the moment, your everyday wants and needs—come to the surface. These thoughts, some generated from Low Ground brain circuitry, others from other impulses and brain centers, have a powerful ability to capture people's attention.

When you move your attention back to your breath, you're active. You don't just wait for your mind to float back; you gently but deliberately move your attention there. This, in itself, is a necessary and powerful way to facilitate moving your mental activity onto the High Ground. You are fighting automatic reflexes by actively guiding your attention. You assertively activate and strengthen the Executive Center of your brain with this mental action. Bringing your attention back to focus is a cardinal example of self-directed neuroplasticity.

Because of this strengthening of the executive brain, your efforts reinforce themselves. With each new episode, it takes less time to become aware of the mind-wandering lapse, and it becomes easier to return. You are actively training your attention density to become stronger and to work more effectively. In so doing, you are training and strengthening your brain. These effects will become increasingly apparent in daily living.[44]

We have found two aspects of applied mindfulness particularly significant: meta-cognition (thinking about your thinking) and meta-attention (paying attention to where you focus your attention). These meta activities are closely interrelated, in that they both involve awareness of and reflection on your own mental processes. We distinguish between them because they have different (though definitely interactive) objects and effects, and it is valuable to consciously pursue them both. (A third meta activity, meta-awareness, is more significant in relation to the Higher Ground.)

Meta-cognition, a term coined by developmental psychologist John H. Flavell in 1979, was first recognized as a way to improve learning.[45] Thinking about your thinking means recognizing your own thoughts and the changes in your thoughts—seeing your own intellectual and emotional progression. Like a novice weight lifter getting used to signals from muscles, you can track the different types of signals sent by and to your brain.

One of the hallmarks of strategic leadership is self-understanding, and thinking about your own thinking is a fundamental part of this.

Are you aware of how and why you arrive at the decisions you make? Do you understand the biases you bring to a decision, and the ways in which your emotions influence your thinking?

Many people don't know much about how they think. They know what they care about and what they find rewarding, but they don't know their own minds or how they arrive at decisions. With meta-cognition, your thinking becomes more familiar territory. You become more consciously aware of the kinds of thoughts you have. The more skilled you become in meta-cognition—thinking about your thoughts, consulting your Wise Advocate about them, watching their pattern through your mind, becoming mindful of them—the easier you will find it to shift away from maladaptive organizational practices. If you continue to deliberately focus your thoughts on the High Ground, after a while it will become second nature.

The other meta-activity fostered by applied mindfulness is meta-attention. You pay attention to where your attention moves as it moves, and you become more aware of the salience of things in your environment. Salience is the quality of being striking and conspicuous, and thus more likely to be noticed. A more salient aspect of your environment is more likely to grab your attention. Examples of salient things include an itch, a loud noise, a bright color, a habitual feeling, a deceptive message, an expression of love, or a reminder of a deadline you haven't met. The practice of meta-attention trains you to recognize them not just for what they are, but for how your attention responds to them. For instance, when a loud noise distracts you, you are aware not just of the noise, but of your response to it. When a deceptive message rises to consciousness (*Other people don't respect me*), you are aware not just of what the message tells you, but of the fact that it's a message.

When you have some proficiency in meta-attention, you are also attuned to your ability to focus your attention more deliberately. When you move your attention back to your breath, for example, you know that this is a deliberate act, and that you have done something valuable. This mindful attention activates and strengthens your executive brain, and thus invokes the High Ground. You actively observe, not passively absorb, what's going on around and inside you.

A growing body of research confirms that these activities of the mind have a significant effect on the brain. For example, Wendy Hasenkamp, currently the science director of the Mind and Life Institute in Massachusetts, conducted research at Emory University on a basic mindful breathing exercise. When people were focusing attention on their breath, fMRI scans showed activation of the lateral prefrontal cortex, linked to the Executive Center function and thus to what we call the High Ground (specifically, in the brain, the dorsal or upper part of the lateral prefrontal cortex was activated). But when their minds inevitably wandered—when they started thinking about the day's activities, obligations, hopes, fears, or anything other than their breathing—the brain scans showed activity in the ventral medial prefrontal cortex, associated with the Low Ground. When they returned to focus on their breathing, as meditators are trained to do, their mental activity returned to what we call the High Ground.[46]

Other aspects of these experiments are also of significant interest. There is a brain area known as the "salience network."[47] It is associated with a part of the brain called the insula cortex (part of the Warning Center), which is related to the "gut brain"—the source of the uneasy but impossible-to-ignore sensation that something demands your attention. When you are in different patterns of mental activity, different phenomena become salient. On the Low Ground, urgent problems capture your attention; on the High Ground, you are more likely to be interested in systems and complex interrelationships.

In Hasenkamp's experiments, when experienced meditators noticed that their mind had wandered, the insular cortex was activated. It appears that through the practice of mindfulness, the subjects had trained their brains to become aware of their own mind wandering. Or more precisely, they had turned mind wandering (which is perhaps the epitome of a nonsalient activity) into a process salient enough that it activated the brain's salience network. That's sort of like a fish training itself to become aware of the water it's swimming in. These individuals no longer had to consciously choose to pay attention to where their attention was focused; meta-attention happened automatically. We view this as a powerful example of self-directed neuroplasticity.

Even more important, the continued practice of mindfulness and contemplation strengthened the High Ground and executive brain connections to the salience areas, enhancing attention control. This continued later, after the session was over, even amidst the crush of nonmeditative everyday life. Finally, continued practice enhances the ability to quiet Low Ground activity and to disengage attention from the Low Ground altogether.

Hasenkamp's research on focused attention, along with other research on mindfulness, has helped explain why these practices are linked both with stress reduction and with increased emotional intelligence. The greater connection between the Executive Center and the emotion-based Warning Center allows more access to attentional control when responding to Warning Center brain messages.[48]

Another researcher who has studied the effects of what we call applied mindfulness on the brain is Norman Farb at the University of Toronto. In one experiment, he asked students who took a basic eight-week course in mindfulness to reflect on words describing personality traits (for example, "melancholy" or "confident").[49] The students used either a mindful perspective, which viewed the experience as an impartial spectator would and did not take it personally, or a more directly personal perspective, in which they ruminated about the relevance of the personality traits to them (Farb called this a "narrative focus"). The same task was also assigned to people with no mindfulness training.

Those with no training in mindfulness had a strong connection between the vMPFC (our Low Road, the Reactive Self-Referencing Center) and the "gut brain" (the insula). The insula, as you may recall, is the source of "gut-level" thinking and feeling in the Warning Center. When the Reactive Self-Referencing Center is strongly connected to the gut brain, the result is often rumination—repetitive, habitual messages that are taken very personally and are often related to stressful situations and the anxiety associated with them. Moreover, those with no mindfulness training had only a very weak connection between their Executive Center and gut brain.

However, for those with just a few weeks of mindfulness practice, connections were significantly strengthened between the insula and the lateral prefrontal cortex (our Executive Center, which is prominent in the High Ground circuit). At the same time, the connection between the Reactive Self-Referencing Center and the gut brain was very markedly weakened. In other words, the gut brain's connection shifted from Low to High Ground. This is very powerful evidence of self-directed neuroplasticity.

Applied mindfulness, like physical exercise, is not necessarily intrinsically pleasant. It may or may not feel good while you are in the practice. Its most beneficial effects are often felt after the practice is over. Though you may not feel as though you have done anything particularly special, the effects are real and significant. They show up on fMRI and similar devices. More important, the practice of applied mindfulness becomes apparent in your ability to focus attention, in your depth of self-awareness, and in your presence as a leader.

Applied Mindfulness and "Free Won't"

If you are an executive with multiple responsibilities, who may not get enough sleep, and who is subject to all the pressures that senior executives face—including a group of people reporting up who may be saying what you want to hear—then you will naturally be drawn to the Low Ground, if only because of all the urgent problems that cross your desk, requiring expedient solutions. Despite its day-to-day usefulness, you may soon come up against the limits of this pattern of mental activity. Applied mindfulness can help you move beyond it and build the skill of shifting to the High Ground.

The neuroscience underlying the effect of applied mindfulness is still a subject of ongoing inquiry and debate. The debate, which is highly significant for strategic leadership, has to do with the nature of free will. On one side are behaviorist and materialist models of the brain, models that held sway for most of the twentieth century and are still prominent today. From this point of view, human behavior is largely determined by physical factors; brain circuitry, the source of messages, is influenced by hormones, physiological factors, or genetic inheritance. Those who think they have free will, or volition over what they choose to think and how they choose to act, are simply responding to material stimuli. Under this model, people are truly trapped by their physiology; there is no point in trying to change brain messages through a discipline of thinking, and there is no way to improve your influence or capacity for self-governance. Leaders are those who are lucky enough to have the right brain-messaging ordained for them.

On the other side is our emerging understanding of self-directed neuroplasticity. When you recognize the degree of choice and intent you have in mental activity, learn how to exercise it, and are willing to do so, then you truly have free will.

Fortunately for those everywhere who would seek to become better leaders, the experiments around "free won't" have demonstrated conclusively that brain messages and impulses do not automatically govern behavior. These experiments began in the mid-1960s, when the German neurophysiologists Hans Kornhuber and Lüder Deecke tracked the pattern of electrical activity in the cerebral cortex just before subjects consciously initiated a movement. They discovered that about half a second before a person moved, a specific electrical brain wave appeared. They called this the "readiness potential," because it seemed related to the person's getting ready to move.[50]

Picking up this research in the early 1980s, neurobiologist Benjamin Libet conducted a highly regarded set of experiments in which he linked the electrical activity to conscious awareness of voluntary action.[51] Tracking brain activity through electrodes, and carefully measuring the timing of how brain signals are related to the awareness of a will or desire to move and to the actual movement of the body that follows, Libet found a consistent sequence, which many others have since replicated. First comes the brain signal, the readiness potential. Then, about three-tenths of a second later, the desire to move arises. This is when people become consciously aware of their inner impulse, for example, to move a finger. Following that by about two-tenths of a second, the voluntary action can take place: the finger can move. But here's the key: it doesn't have to.[52]

Libet's experiment is particularly relevant to the deliberate restraint involved on the High Ground. Is it possible to resist the impulses and emotions of the brain, the habitual responses of deceptive brain messages, or the continual transactional pull of the Low Ground? Not only is this possible, but, according to Libet and subsequent research, because of the delays between the brain signal, the inner impulse (or desire), and voluntary or willful action, potential moments of awareness and restraint are continually available to you. Various experimenters, using much more sophisticated equipment than Libet first used in 1983, have found time lags in which brain signals occur more than ten seconds before the desire to move comes into awareness. And as we said, *the finger doesn't have to move.* During the brief delay— about two-tenths of a second after people become aware of the inner impulse, just as Libet found—they can choose to either follow it or not. Moreover, because decisions to inhibit movement (which neuroscientists call exercising inhibitory control) are generally associated with the Executive Center in the brain, there is reason to think the High Ground is involved in this process of inhibiting a response.[53]

The four-step process described in this book—and particularly the first step, relabeling, described in chapter 4—provides a practice for making this choice about the impulse and exercising inhibition. It gives you power where you might otherwise be powerless. Once the brain signal causing an inner impulse (the deceptive message) occurs, you cannot prevent it from reaching your awareness. You cannot control desires or impulses or prevent them from "bubbling up," as Libet refers to it in a 1999 paper reviewing his own long-standing work on this subject.[54] Nor should you try to control these inner impulses, at least not in the short run. It's perfectly "human" and normal to have feelings, cravings, and impulses of all sorts and the desire to follow them. Human brains are wired for this, perhaps in a way that evolved from the need to continually look out for potential threats. This means that, on some level, people don't have full free will; they cannot prevent a thought or impulse from arising in the short term.

Does that mean that you have no choice except to act in accordance with your impulses? The answer is clear from the same experiments—and probably from your own life experience. You have the power to identify this impulse or desire as an impulse, and then *not to act* on it. Libet gave the name "veto power" to this choice, but the more common (and very evocative) name for it, attributed to psychologist Richard Gregory, is "free won't."[55]

Free won't is the ability to recognize brain-based cravings without giving in to them or acting on them. This is a primary executive function (also known as inhibitory control)[56] and thus closely related to the High Ground pattern of mental activity. Anyone can strengthen this ability through applied mindfulness and the practices described in this book (relabeling, reframing and refocusing). When the ability is well-developed, then at any time, you can stop your brain messages, deceptive or otherwise, from dictating your behavior. Importantly, the exercise of free won't is not a matter of sheer brute willpower; you are not suppressing your impulses. It is a matter of awareness. The practice of relabeling makes you more aware of your inner impulses. It helps you to recognize an impulse as just one of your brain messages. Once you recognize it, you can then decide whether or not to respond to it.

You can also use the Wise Advocate to help master this process. By actively saying to yourself, *I'm going to consult the voice in my mind, the voice that looks out for me over the long run*, you trigger a pattern of mental activity that strengthens your ability to restrain your impulses.

Doing this type of thing over and over again rewires your brain, which over the medium and long term actually does give you more access to your executive function and thus more control over your inner impulses. Self-directed neuroplasticity, changing your own brain, can change and decrease the power of deceptive brain messages, and the impulses they cause.

This exercise of "free won't" is one of the definitive things that separates human beings from animals. To exercise this choice, you need to be aware, in the moment, that there are alternatives to the instinctual, habitual response that comes naturally to you. The more practiced you are at relabeling, the more conscious you become of your brain's operations, and thus the easier it is to make an adaptive choice, even in an uncomfortable direction. That's the essence of strategic leadership.

The exercise of free won't at any time of day is much easier if contemplative practice is part of your life, even for a few minutes each day. One of the benefits of applied mindfulness is that it strengthens the executive function, making it easier to see the power of your impulses and to let them go. With recurring, disciplined attention to the nature and quality of your own thoughts, feelings, and decisions, you become more adept at connecting with the Wise Advocate. To those unfamiliar with it, contemplative practice may not seem at first as if it's having much effect. But the discipline reinforces itself and builds your mental proficiency—your capacity for developing creative narratives to clarify decision making and enhance your executive capacity.

Approaching Higher Ground

Imagine that you are a leader who has cultivated the High Ground in your mind. You have a busy schedule, and, like every leader of a major enterprise, you are under an immense amount of pressure, but you are disciplined about the way you approach your own development. Every day, for five to ten minutes or perhaps more, you sit in mindfulness practice, simply focused on your breath. You learn to pay close attention to the effects that your emotions have on you, and how distracting they can be, recognizing when you are anxious or under pressure. By refocusing on your breath when your mind wanders, you have gained inhibitory control and enhanced your cognitive flexibility and working memory (the capacity for making complex comparisons and contrasts). You are familiar with the experience of exercising "free won't," and you have a relatively high level of agency with your impulses; you know how to manage them without having to act on them.

You are also reasonably practiced in mentalizing. When someone in your orbit has a problem, you are prone to think about what they are thinking and what they will do next, not just what they want and how to give it to them. You are more aware of the narrative of the Wise Advocate and its relationship to your true self, and the fact that you can hold your own in a responsible position with people reporting to you and tough challenges.

As you combine applied mindfulness with mentalizing and executive-function development this way, all three practices become routine for you. You ruminate less on the expedient interests in your life: *What's in it for me? How will I get out of this mess? How can I get what I want?* And you pay less attention to the questions typically associated with authority: *How will we fix this problem? Who can we bring on board with us? How will I triumph?*

Instead, you apply your proficiency in mentalizing—in thinking more clearly about other people—to yourself. *What am I thinking? What am I likely to do next? What am I really about?* You can thus see yourself, your thoughts, and your patterns of mental activity as Adam Smith's impartial spectator might see you. In effect, you begin to mentalize about yourself—including your own situation and your role in influencing it. *How do I expect to lead?*

Now, when you engage your Wise Advocate, you're creating a narrative about yourself and the system around you in which you play the role of strategic leader. You pay more attention to the questions associated with strategic leadership: *What does this situation call for? What is the best thing to do? Why am I drawn to do this? What will the consequences be?* Questions like these deepen your connection to and dialogue with the Wise Advocate in your mind. Now the High Ground is a transitional path leading to the Higher Ground in your mind.

The High Ground and the Higher Ground are two distinct patterns of mental activity—different enough to be associated with separate behaviors and separate aspects of the physical brain—but they have several key activities in common, including meta-cognition (thinking about your thinking) and meta-attention (paying attention to the way you pay attention). The impartial spectator activity, "mentalizing about yourself," is also common to both states. Those are shown in table 3.2, in the square at middle right, labeled "Applied Mindfulness" (in boldface type). As you become accustomed to the mental activity of the High Ground, you will move up through this square: mentalizing about yourself, then consulting the impartial spectator, and then increasing your practice of meta-cognition (thinking about your thinking) and meta-attention (paying attention to where your attention is focused). This will continue as you accustom yourself to the Higher Ground.

TABLE 3.2
The Boundary Between the High Ground and the Higher Ground

Pattern	Leadership	The Wise Advocate	Activities that Invoke this Pattern	Mental Narrative
Higher Ground	**Transformational leadership:** Making the most of your organization's potential and helping it become what it is called to be.	You consciously and habitually consult your Wise Advocate and play the role of the Wise Advocate for the larger organization, all in the process of developing your true self.	**Deep mindfulness:** Meta-awareness	Am I consulting my Wise Advocate? How would the loving guide in my mind regard what I am thinking and doing? What does it suggest?
			Meta-cognition and Meta-attention	Thinking about what you're thinking; Paying attention to what you're paying attention to. What am I thinking? What am I paying attention to?
			Applied mindfulness: (*associated with High and Higher Ground*) Consulting the Impartial Spectator Mentalizing about yourself	How would an objective, highly credible observer regard what I am thinking and doing? What am I thinking? What am I likely to do next?
High Ground	**Strategic leadership:** Addressing problems that are too complex for an organization to manage easily; helping the organization transcend its limits; being able to rapidly move between High and Low Ground.)	You tend to consult your Wise Advocate when you make important decisions and manage others.	**Executive function:** Cognitive flexibility Working memory Inhibitory control (Free won't)	What does this look like if viewed from a different perspective? How does this relate to what I already know? Should I refrain from acting on this impulse?
			Mentalizing	What is he or she thinking? What are they thinking? What are they going to do?

What, then, is different about the Higher Ground? The key new practice is meta-awareness: cultivating a continuous, sharply honed awareness of your own awareness. This includes awareness of the connection with your Wise Advocate. When you ask *Am I in touch with the Wise Advocate right now?* the answer gives you insight into the ways in which you, personally, handle the most fundamental challenges you face. This could be called deep mindfulness. It generally requires a great deal of practice to consistently reach this state. For many people, it has a profoundly spiritual aspect, and it always engenders a nonmaterial sense of connection with larger systems around you.

Over time, as you inhabit the Higher Ground this way, you create stronger habitual links between the mental activities of your Executive Center (associated with the lateral prefrontal cortex), your Deliberate Self-Referencing Center, and your Habit Center. You no longer need to consciously remind yourself to turn to the Wise Advocate for guidance; it is continually present. Planning, goal-directed behavior, cognitive flexibility, and a broad, objective view of your own actions become embedded in your routines of thought and action. In brain terms, we sometimes think of this as going "around the bend"—because the dorsolateral prefrontal cortex is located in the brain around an inverted U-shaped bend to the side of the dorsomedial prefrontal cortex. The steadiness of the input of the Wise Advocate has enabled a confluence of executive function and mentalizing, of the Executive and Deliberative Self-Referencing centers.

The link between the Executive Center and the Habit Center is not strong for most people, but once it is strengthened, it is second nature to step back and look at your own situation with a much better frame of mind. Psychologists who have studied this state at length refer to it as "intrinsic mindfulness."[57] It is easier to manage stress, and you are less likely to become overwhelmed by Lower Ground mechanisms—or, to be more precise, your executive function is better protected from the deleterious effects of stress, such as the amygdala hijack.

In subsequent research, Farb and others have found that mindfulness practice continues to activate and develop the capacity for inner awareness.[58] More proficient mindfulness practitioners have more control over these connections than the High Ground offers; they can turn up and turn down awareness of their gut feelings.[59] It's as if they have a rheostat in their brain, dialing up or down their awareness of their emotions and of every other aspect of the present moment.[60]

You observe what is happening, you experience it, you deliberate and evaluate it, and then if you so choose, you can let it go.

The Higher Ground requires people to transcend the animal aspects of human nature. This is traditionally a religious goal, not a secular one. Secular groups, including nearly every major business organization, are not equipped to ask people to make that type of commitment. Arguably, they should not be required to do so; it is not fair to ask people to make a religious commitment as a requirement for their livelihood.

And yet there is reason to believe that the value of spiritual commitment—or at least the frame of mind resembling spiritual commitment—is significant for strategic leadership. In one set of experiments by the Chinese researcher Shuihui Han at the University of Beijing, people with a strong Christian perspective were found to gravitate to the Higher Ground and its High-Ground-associated brain areas. (A close examination of figure 3.2 shows an overlap region between the High and Higher Ground.)

The experiment compared two groups of people: a group of control subjects who were not heavily motivated by religion, and a group of devout people who actively said that Jesus Christ was significant in their inner life.[61] Both groups were asked questions that typically trigger the Self-Referencing Center; as you recall from chapter 2, this is the part of the brain (the medial prefrontal cortex) associated with the concept of self. To engage their sense of self, participants were asked to take a test in which, one by one, they would hear words in their native language (Chinese) and say whether they felt more or less identified by them, compared to a well-known Chinese personality used as a reference. Are you an actor? A comedian? A basketball player? Are you healthy? Muscular? Liberal? Conservative? Hospitable? Idealistic? Angry? Happy? Prompt? Trusting? Perfectionist? Spontaneous? Imaginative? Improvisational? A total of 240 words were selected based on generally accepted research indicating that they were strongly connected to people's sense of self.[62] It didn't matter so much what choices the subjects made; simply by considering the question, they were activating their Self-Referencing Centers. But which part of the medial prefrontal cortex would be engaged—the ventral (lower) section (the Reactive Self-Referencing Center), associated with the Low Ground, or the dorsal (upper) section (the Deliberative Self-Referencing Center), associated with the High and Higher Ground?

When given these tests, the devout Christians were much more strongly engaged with the dorsal medial prefrontal cortex, the Deliberative

Self-Referencing Center. Han found a similar effect with Buddhists who regularly practiced meditation.[63] On the other hand, those who had no systematic faith were much more likely to activate the ventral section associated with the Reactive Self-Referencing Center and thus to stay on the Low Ground. One correlation was particularly pronounced—between activation of the dorsal medial prefrontal cortex (High Ground brain functions) and assessment of "the importance of Jesus' judgment in subjective evaluation of a person's personality." Han concluded that what we call the Deliberative Self-Referencing Center was evoked through devotion to something that transcended the self: "Christianity strongly encourages its believers to surrender to God and judge the self from God's perspective."

Whether you associate it with any particular religion or not, there is clearly a link between the Higher Ground and a deeper sense of meaning. Those who do not simply let their thoughts pass by without any transcendent context will be drawn to think about fulfilling themselves with a purpose greater than themselves. When you become accustomed to thinking that there is a perspective higher than your Low Ground sense of self, it makes the Deliberative Self-Referencing Center and the associated dorsal area stronger. It weakens the perennial reactive tendency to ask *What's in it for me?* and strengthens the ability to consider *What am I being called to do?*

Leading With Meta-Awareness

In our opinion, the leaders who matter the most—the transformational leaders who make the greatest difference—are motivated and animated by patterns of thinking that continually move them toward the Higher Ground. Their habits of mentalizing and mindfulness— "mentalizing about themselves"—represent a deep form of humility; it is an approach that allows them to think about their own thinking from the perspective of something outside themselves.

Meta-cognition, meta-attention, and meta-awareness are all linked together, reinforcing one another; they represent the felt experience of a growing level of attention density. Researchers have clearly related this type of profound mindfulness to the awakening of the brain's Executive Center—the prefrontal cortex. It also seems to enhance the capacity for empathy and clear-headed decision making, even as the neural grid associated with the Higher Ground becomes more complex and interconnected.

Even before you get to the Higher Ground, you'll find that being a strategic leader operating regularly on the High Ground creates new habits of mental activity for you. As you practice applied mindfulness and gain skill, the Wise Advocate takes shape within you and becomes, if not a literal presence, a source of steady, loving guidance that you can regularly turn to, and emulate as well. This enables you to play a broader Wise Advocate role for your organization. As you become increasingly aware of the Wise Advocate's perspective, and especially of its presence in your mind, then you are increasingly ready to take on the role of a Wise Advocate for others around you.

Leaders on this path will move beyond purely strategic considerations and become true transformational leaders: dedicated to helping the organization (and potentially the community around it) make the most of its potential. If you play this role, then you are continually listening for what the organization is called to become, and helping it step into that future. If enough of your colleagues, throughout your organization, can rise with you to similar levels of meta-awareness, then your organization might become to society what you are to the organization: a Wise Advocate, actively engaging with the larger system around it.

To paraphrase an aphorism attributed to Margaret Mead, never doubt that a small group of people consulting their Wise Advocates can change the world for the better. In fact, that's the only thing that ever has. In the rest of this book, we'll show you how.

Reflection

How accustomed are you to mentalizing? For example, how much do you think about who your customers are and what they're thinking about? About who your employees are and what they're thinking about? Or about who other stakeholders may be and what they are thinking?

Do your consult or hear the voice and encouragement of your Wise Advocate when making decisions?

What role does mindfulness play in your current decision-making processes?

What benefits could applied and deep mindfulness deliver for you, your organization, and your results?

Are you accustomed to thinking about your thinking? Do you pay attention to the way you pay attention? Are you aware of when you are aware?

CHAPTER FOUR

Relabeling Your Messages

DO YOU REMEMBER the 1944 film *Gaslight?* Adapted by filmmaker George Cukor from a 1938 play by Patrick Hamilton, it concerns a nineteenth-century murderous husband (Charles Boyer) who tries to convince his young bride (Ingrid Bergman) that she is insane. He hides jewels and other valuable objects and accuses her of misplacing or stealing them. He manipulates the gas lighting to make it flicker, thus convincing her that she is going mad. When she accuses him of lying to her, he says she must be delusional.

Her own responses, based on the assumption that her husband's view is accurate, add to her confusion and vulnerability. She withdraws from the world, telling herself that he is right and there is no hope for her but to be committed to an asylum. But when her husband is caught and exposed as a thief and a murderer, she realizes the false nature of the messages he had been sending her way. Ever since that film appeared, the word "gaslighting" has been used to mean manipulating someone else by pretending that deceptive messages are true.

By far the most common kind of gaslighting, however, is the kind that people do to themselves. If you're like most people, thousands of automatic messages course through your brain circuits every day, coloring your perception of the world. Generated in daily thought and conversation, they undermine your abilities as a strategic leader. When you experience a recurring rumination—*I always get this wrong*, or *Nobody appreciates me*, or conversely, *I'm so special I can get away with anything*, or *Everyone else sees things the same way I do*—you are most likely experiencing a signal generated by your own brain circuits. These thoughts tend to distract or dissuade you from your most important goals and intentions, but they seem so natural that most people tend to accept them as reality or feel them to be irresistible. Doubts about them can feel like delusions. They also affect the way you pay attention to the world, making you more likely to notice events and phenomena that reinforce your habitual thoughts. They thus continually lead you to confirm their own skewed logic—which is why much bias is known as "confirmation bias."

To be sure, some of these deceptive brain messages contain elements of truth, or they wouldn't be convincing. But as we saw in chapter 1, we call them "deceptive" because they are automatic and intuitive, and thus linked to the Low Ground pattern of mental activity. This makes them innately detrimental to your best interests, at least in the sense of masking more fundamental, strategic opportunities for you and your organization. Most of them reflect some past experience that is no longer as relevant; the world has changed since your brain adopted them. If you treat them as self-evident and heed them, you shortchange your own potential.

By adulthood, if you're typical of most people, you have learned to manage your own deceptive brain messages to some extent. You know that you must sometimes step outside of your comfort zone to learn and achieve new things. For those who cannot do this—those with problematic conditions like obsessive-compulsive disorder and some forms of addiction and depression—relabeling is often the beginning of therapeutic relief. They need to become aware that these are simply brain messages, not reality.

But even those of us who have reached maturity need to practice relabeling, especially when we begin working for organizations. Many people in organizations see a wide gap between the role they would like to play—the way they would prefer to run their life and their part of the business—and reality as it exists. They would like to spend less time on trivialities, to act with more autonomy, and to focus their attention on the issues that matter. But if you ask what's stopping them from doing this, most of the barriers come from their own assumptions, for example: *If I try to change the way things work around here, my career will suffer.* Assumptions like these, crystallized into deceptive organizational messages that become embedded in the company's culture and manifest in everyday conversation, make it more difficult to remain on the High Ground. Relabeling deceptive messages is critical for moving toward strategic leadership.

We call this step "*re*labeling," rather than just "labeling," because deceptive brain and organizational messages already have labels. They are labeled "my thoughts" or, more simply, "the reality I know" or "the truth." When you relabel them, you assign them to a new category: messages. *They are not reality. They are not me. They are simply manifestations of my brain's physical activity. There is a constant flow of these deceptive messages, and I can watch them going by.*

The experience of relabeling often involves consulting your Wise Advocate, and it brings the High Ground into play. That helps break the Low Ground habit of treating them as reality. The Wise Advocate, as a construct with your best interests in mind, confirms that these messages are not accurate. They are not coming from your essential self. If they were, they would lead you in a different direction.

To fully understand the power of relabeling, you need to know what is happening in the Low Ground brain circuit when you receive deceptive messages. As you may recall, this circuit links your Reactive Self-Referencing Center ("who I am, what I want, and what others want") with your Warning Center ("something is wrong") and your Habit Center ("do what feels right.") Deceptive messages, whether pleasant or painful, flow easily because they are habitual.

Relabeling does not mean suppressing these brain-based messages and impulses. Instead of trying to pretend they don't exist, you acknowledge them, examine them more closely, and cultivate your awareness of the way they circulate in your brain. (The organizational learning theorists Chris Argyris and Donald Schön referred to this type of reflection as double-loop learning.)[1]

You may be wondering if you should replace them right away with new messages, a process we call "reframing." That is the next step, described in chapter 5. We have learned to keep relabeling and reframing separate, because they both require concerted effort and concentration. Relabeling is more effective when your attention is fully devoted to it. It takes real attention, and often consultation with your Wise Advocate, to come to a full sense of the nature of these messages and to recognize your power over them and your ability to resist them.

In relabeling, you put "free won't" into practice. Free won't, as you may recall from chapter 3, is the human ability to resist impulses without denying that they exist. In the step of relabeling, you see the impulse as just an impulse—one you don't need to obey, at least not right now. Thus, when you are provoked to lash out in anger or frustration, you can step back and relabel the sensation: *This is the feeling I always get when I have an angry deceptive brain message.* Similarly, when you are about to do something addictive, you can relabel the impulse as the craving you get from your brain, and choose to say no to it—at least just this time. You may of course choose to indulge the sensation or the craving, but in terms of the neural dynamics of your mind and brain, it is a choice. Making that choice more consciously tends to lead your mind, and your brain, toward the High Ground rather than the Low Ground.

Gaining proficiency in this type of neuroplasticity may not come easily; for many people, it is a lifelong struggle. You may find that the relabeling itself—the thought that this impulse is just a deceptive brain message—generates other deceptive brain messages. They may include a rabid anger at the very idea of relabeling in the first place. *What am I supposed to do with all this frustration? Ignore it?* Or, *I don't have time to slow down and think differently about this.* Or, *Other people find it easier to relabel than I do. They don't get what I'm feeling right now.* Or, *If I don't give in to my impulses, then I'm just surrendering to the outside world that wants me to suppress myself.* Or, *I'll relabel if you relabel first.* Those, too, are deceptive messages from your brain. Relabeling them as such changes your reaction from something that compels you to something you choose.

You'll probably see some results from relabeling fairly quickly. It will be easier to manage deceptive and difficult impulses, feelings, and messages. But you will also undoubtedly relapse. This is a lifelong task. Over time you continue to develop the skill of inner observation, taking heed of your conscious experience and making better choices as a result.

The act of relabeling may seem simple and rather paltry, but it is one of the most powerful things you can do as an individual, especially when you do it over and over. It is an application of applied mindfulness and, more specifically, of metacognition—thinking about your thinking. You become less likely to speak out in anger, to betray a secret, or to give in to a bad habit or an addiction. By abandoning the automatic assumption that these messages are accurate, you assert the agency of your mind.

Lauren and Majid Relabel

Consider, for example, the case of Lauren, a regional operations manager for a rapidly growing family-owned specialty beverage company. The company's best customers are people who prefer premium, painstakingly sourced coffees and teas. Lauren had looked forward to a blossoming career—until Majid, the warehouse manager for the same region, started routinely going over her head to "clarify" her decisions, as he called it, by asking their bosses, the senior leaders of the company, for approval. The problem came to a head when he questioned a decision that Lauren had made to postpone expansion to a new territory, a group of small cities that Majid was already setting up the warehouse to serve. She had asked him to hold off for another six months because she was still researching the market, and he had gone to the bosses to protest.

From Lauren's perspective, this was problematic on several levels. It forced the senior team at the head office to reassess a relatively minor decision that she had already made. It made Lauren, who outranked Majid, look like someone who wasn't in control. And it confused the staff in the region, who were subordinate to both Lauren and Majid. Suddenly, they couldn't be sure who to check with when they needed guidance on a thorny issue.

The harder Lauren struggled to unravel the problem, the more difficult it became. It got to the point where she couldn't keep from thinking about it, even during her time off. Her running internal commentary sounded something like this:

Why is Majid doing this? He's from the Middle East. Maybe he doesn't want to work with women. Or is it just that he doesn't like me? He seems to have no faith in my decisions. In fact, nobody respects my decisions; if they did, I wouldn't have this problem. I'm not capable of being a leader in this company. Something must be wrong with me.

These messages were hard for Lauren to suppress or ignore, because any possible threat to her job strongly triggered her Warning Center. (As you may recall from chapter 2, this is the brain area containing the amygdala, the insula, and the orbital frontal cortex, all associated with the sensation that "something is wrong," involved with both Low and High Ground.) The more she ruminated about it, the more she seemed to be at risk. Her skills, personality, and values were completely in tune with this company. She had never enjoyed working anywhere as much. She liked the founders and their commitment to local community involvement, and her innovative work on customer service had been a recognized factor in the company's growth. Yet if things continued the way they were going with Majid, she would be forced to leave the company. She started to withdraw from conversations with all her coworkers, she no longer spoke up in meetings with the founding family, and she quietly began to circulate her resumé.

Lauren was caught in a Low Ground trap. Fortunately, she wasn't all-consumed by her deceptive brain messages (few people are). She sensed that the situation had been blown up out of proportion. If she could just talk to Majid openly, she could solve all this.

Thus, after some reflection, Lauren decided to raise her concerns with someone she knew at work, someone who had acted as a mentor to her. This individual saw the situation the way a Wise Advocate might: Lauren was operating from assumptions that hadn't been tested and might be misleading her. The mentor advised Lauren to screw up her courage and ask Majid to have lunch with her, just the two of them.

Asking Majid to lunch wasn't easy; and when they met, speaking openly about her concerns was even harder. But she did it. "Majid," she said gently and without rancor, "we have a problem between us, and it's affecting everyone on our team." And Majid, to her surprise, opened up. He said he was just as concerned about the situation as she was. He worried, he said, that he was seen as a whiner and a snitch. But he felt he had no choice but to speak up. When he was given his current position, he had been told to check in with headquarters routinely whenever he saw a problem. If he didn't do that, and they found out about these problems separately, he would lose credibility.

Majid, of course, had deceptive brain messages coming from his own Low Ground circuit, and they ran something like this:

Lauren is jeopardizing the company by being too cautious, and I'm the only one who sees it. Meanwhile, the company founders need me to keep the operations going, and they'll find out from my staff if I miss anything. They clearly like Lauren better than they like me, because I am from the Middle East. Clearly, I can't trust anyone here. If something goes wrong, I'll be the first one to be blamed.

Having opened up a little bit about their suspicions, they now cautiously moved to discuss the main thing they had in common: their interest in promoting growth for the company in this region. From that perspective, they were natural allies.

Lauren realized that Majid, rather than wishing to undermine and control her, was simply trying to protect himself. She didn't have to start trusting him wholeheartedly, but she did have to give him enough benefit of the doubt to ask his opinion whenever she saw a problem.

Majid, for his part, recognized that Lauren had the founders' confidence—not because of her ethnic background, but because of the way she worked with them. He realized that he could learn a great deal from her.

A few days later, Majid asked his bosses to clarify the role they wanted Lauren to play at the warehouse. They surprised him by saying that they didn't want her to play any role there. That was his job. But they expected him to work closely with her and develop the kind of operations that would allow her to expand their market.

In other words, Lauren and Majid realized, their success depended on each other. They began to talk more openly together and found new ways to collaborate on expanding their business. Each has since come to regard the other as someone to rely on. Relabeling their perceptions—realizing that their views of each other were just brain messages that misled them—enabled them to do this.

Relabeling can help prevent an organizational train wreck that would otherwise be unavoidable. If Lauren had allowed her deceptive brain messages to prevail—if she continued to tell herself that people were trying to undermine her, or that other people were likely to outshine her—then she would have reacted accordingly. In effect, she would have blocked her own career and effectiveness, and diminished her value to others as a leader. The consequences would have been bad not only for Lauren, but also for the organization: her contribution would be lost. Similarly, if Majid had allowed himself to continue to be governed by his deceptive brain messages, he would no longer have been effective in his job.

Deceptive Organizational Messages

Many organizations have a larger-scale problem analogous to deceptive brain messages. A constant flow of deceptive organizational messages leads the organization in dispiriting, unprofitable directions, usually based on inaccurate perceptions.

Deceptive organizational messages can be even more powerful than their individual-brain counterparts, because collective conversation keeps them active. They become part of the workplace culture, the "way we do things around here." Indeed, when company leaders complain about their culture, they're usually complaining about these corporate cognitive distortions. When a leader says about a proposed idea, "We tried that in the past and it didn't work," an implicit consensus often follows: *It will never work and never could.* People treat this message as an unquestionable axiom, assume that others believe it, repeat it up and down the enterprise, and avoid any action that would contradict it.

In cases of repeated sexual harassment or other severely counterproductive behavior, an organizational deceptive message often contributes to collective denial: *That's not the kind of company we are. Therefore, this must not be happening.* Even when people try to ignore these messages, they affect the enterprise. As Stanford researcher James Gross has noted, the act of resisting or suppressing deceptive brain messages leads to a higher level of stress for individuals. The same is true for groups.

The organizational theorist Chris Argyris apparently had deceptive organizational messages in mind when he wrote about "undiscussables"— attitudes and beliefs that guide what people do, but that don't get talked about openly. People feel they can't talk about them, because they don't have permission. Nor can they talk about the fact that they can't talk about them. So they remain silent, which simply reinforces the messages. In that type of organization, few people feel that they have permission to speak out constructively, as a Wise Advocate or impartial spectator would. Instead, people act as if these deceptive brain messages are accurate statements about reality. It's as if everyone in the company is "gaslighting" one another—deceiving themselves and their colleagues about the range of options available to them and their own potential effectiveness.[2]

When an organization loses its way—when it fails to keep innovating or misses a strategic opportunity, for example—you can almost always bet that people, as a group, have embraced some deceptive brain messages, resulting in corporate cognitive distortions (deceptive organizational messages), and are repeating them to each other. One ongoing task of a strategic leader is to bring deceptive organizational messages to the surface and to help people recognize them as merely messages, not automatic reality. Once they are relabeled this way, they lose much of their power. It takes a powerful, strategically minded leader to accomplish this. You have to stand back from your everyday transactional world and recognize those messages yourself before you can communicate that awareness to others.

It may help to know that deceptive organizational messages tend to be generic; the same ones are active in many different organizations. Based on our experience and observation, and on some of the observations in *You Are Not Your Brain*, we think there are at least four broad categories of inaccurate misperceptions that you are likely to encounter (see table 4.1 for a summary). Each has its own pernicious effects, but each, with the right kind of leadership, can be relabeled and reframed.

TABLE 4.1
Common Organizational Deceptive Messages

Misperceptions		
Risk	Overconfident exceptionalism (complacency): *The dangers don't apply to us.*	Excessive risk aversion (catastrophizing): *We must prevent, or at least prepare for, every unpleasant possibility.*
Value	Perfectionism (all-or-nothing thinking): *It must be flawless or it's worthless.*	"Ticking the box": *If it meets the specs, it's good enough.*
Proficiency	Entrenched insecurity: *We are not effective and never will be.*	High self-efficacy (mind reading): *We are so effective that everyone agrees with us.*
Validity	Emotional reasoning: *If we feel good about it, it must be true.*	Rigid rationalism: *We came to this decision rationally, so there will be no disagreement.*

1. Misperceptions of Risk

"Again and again," note economists Carmen Reinhart and Kenneth Rogoff, "countries, banks, individuals and firms take on excessive debt in good times without enough awareness of the risks that will follow when an inevitable recession hits." The title of their book, *This Time Is Different*, is a reference to the deceptive message voiced during the buildup to the financial crisis of 2008—and before similar crises throughout history: "We are doing things better, we are smarter, we have learned from past mistakes," write Reinhart and Rogoff, paraphrasing mistaken assessments of risk. "The old rules of valuation no longer apply."[3] In these statements, "we" generally refers to the financial establishment, which is falsely credited with finally figuring out how to design a crash-proof economic system.

Overconfident exceptionalism of this sort, in which executives underestimate the riskiness of their activity, has led many companies to complacency, and then to failure. *We don't have to worry about losing customers*, executives say when faced with an upstart competitor. *They have nowhere else to go.* Sometimes this type of deceptive message arises around a narcissistically heroic leader. *Our CEO takes chances that no one else could get away with.* If the exceptionalism extends to the entire company, then managers get into the habit of overstepping boundaries or fudging numbers, growing bolder and bolder until the risks catch up with them. *The dangers don't apply to us.*

The flip side of overconfident exceptionalism is excessive risk aversion. This can be equally debilitating, especially when it becomes habituated as a way of life. It is known in psychology as "catastrophizing"—ruminating about worst-case scenarios. *We must prevent—or at least prepare for—every unpleasant possibility.* Excessive risk aversion often takes the form of accumulating as much support for a decision as possible before approving it, either from data (a practice known as "analysis paralysis") or from colleagues and overseers within the company. *It looks OK to me, but we can't take the chance. You'd better ask other people for their approval as well.* It leads decision makers to shut down entrepreneurial decisions and forego valuable opportunities—including the opportunity to learn from risky situations and build up their own capacity for judgment. Excessively risk-averse companies unintentionally take the greatest risk of all: being left behind because they can't make decisions rapidly enough.

These two deceptive messages can coexist. An organization can swing from one to the other. Underlying both is the same perception: that the decision maker's comfort level is an accurate indicator of risk. Actually, comfort levels are often problematic indicators; they are derived from past experience with success (that might not continue) or painful failure (that need not happen again). Though the skill of risk assessment is fundamental to strategy, it is difficult to develop in the face of these deceptive organizational messages, especially when they aren't recognized as such.

2. Misperceptions of Value

These deceptive messages provide a misleading idea of the worth of current human endeavor. People judge their own (and each other's) activities as high- or low-potential based on habitual reactions. Often these misperceptions manifest as perfectionism, or all-or-nothing thinking: *It should be completely flawless, or it won't be worth anything.* Under the influence of this type of message, a functional team might decide not to propose an interesting idea because they fear it isn't good enough. A research group might second-guess an innovation, drag it down with extra features, and delay it until it's eclipsed by rival offerings. A supervisor, considering promotions for the staff, might oscillate between extremes: treating a direct report as a star one year, but deeming that person a total screw-up the next.

The opposite of all-or-nothing thinking is "ticking the box": accepting suboptimal work as long as it complies with specifications. *It's close enough for government work* is a deceptive message of this sort. *It doesn't matter how good it is, as long as we meet the deadline* is another. This type of message leads people to underpromise so that they can underdeliver without penalty, to dismiss improvement efforts as not worth the cost, and to look the other way when their colleagues cut corners.

Misperceptions of value often reflect the "fixed mind-set" perspective identified by Carol Dweck. If people's basic worth is fixed in place by the time they come of age, limited by the talent, intelligence, and circumstances they have inherited or acquired as children, then static judgments of value make sense. As Dweck points out, a more accurate view is the "growth mind-set"—the idea that people can change habits, transcend limits, and expand their capabilities throughout their lives. Indeed, people continually do this through self-directed neuroplasticity. They focus their attention, over and over, in a way that builds new habits by etching new neural pathways in the brain. If you believe in the growth mind-set, then neither all-or-nothing perfectionism nor "ticking the box" makes sense; instead, you regard human activity as an investment worth making if it will lead to genuine learning and consistently improved results.[4]

3. Misperceptions of Proficiency

How capable are you and your company of influencing others and getting things done? Your answer reveals an attitude that psychologist Albert Bandura termed self-efficacy.[5] Self-efficacy is confidence in one's own ability to succeed. People with unrealistically high self-efficacy assume they will prevail at difficult tasks even if they lack the proficiency to do so. People with excessively low self-efficacy are likely to give up even when they could actually succeed. Deceptive organizational messages can carry either misperception.

In organizations, low self-efficacy manifests as entrenched insecurity. Entire groups internalize the idea that *We are not effective now, and we never will be.* In some hospitals, for example, nurses as a group are reluctant to volunteer information to physicians, even when they see a patient's condition deteriorating. They feel they don't have the status or expertise to be heard, even though they are often the most expert people about patient care, comfort, and oversight. Patients have died because nurses felt they could not bring something they saw to a doctor's attention. This misperception often involves the cognitive distortion called "discounting the positive": any good attribution to your company or your work must be false.

One of the most damaging aspects of entrenched insecurity is the way it leads people to misinterpret other people's signals, as Lauren and Majid did. Both Lauren and Majid assumed that they were each other's targets. When Majid went around Lauren, she interpreted it as a sign that she was a failure. When Lauren asked Majid to hold back his expansion plan, he assumed it was an attack. Underlying these assumptions were similar deceptive messages. Unknown to each other, both felt they would never be fully accepted, Lauren because of her gender and Majid because of his background as a foreign national. Meanwhile, their bosses had seen both of them as high-potential managers—until it began to seem as though they couldn't work well together.

The other common misperception of proficiency is excessively high self-efficacy. This tends to take the form of "mind reading": projecting your own attitudes onto others, assuming that they share your opinion about yourself and the situation, and acting on that assumption. *We are so effective that everyone agrees with us.* This type of deceptive message is prevalent in technology companies. Software specialists assume that a complex user interface, because it intrigues them, will similarly delight their nontechnical users. The engineers discount the novices' complaints. *When they get used to this system, they'll appreciate it.* In other companies, mind reading leads to underestimating customer concerns, for example about privacy or security. *Of course they trust us.* Mind reading is also at play in many cases of workplace sexual harassment or other abusive behavior. *When people say no to me, they don't really mean it.*

4. Misperceptions of Validity

Misperceptions of validity lead us to believe that something is true, and will be treated as true, either because it feels right or because it is supported by a logical argument. These cognitive distortions split reason from emotion, even though the most effective, long-lasting decisions bring together both types of validity.

One common type of misperception of validity is known as "emotional reasoning." This deceptive message suggests that, because you and your colleagues feel something is true, it must therefore be true. When you judge the logic of a decision based on how it makes you and your colleagues feel, you may be led astray. This pattern often affects deals, because people tend to evaluate investments based on their emotional impression of past transactions: *We were stung by our last deal in this region. Never again.* Or, conversely, *The last time we made a deal with that group, it went terrifically. We won't need to worry about this one.*

One prominent example of emotional reasoning is loss aversion, a dynamic identified by Amos Tversky and Daniel Kahnemann. People have a stronger reaction to the prospect of losing something than of gaining something, even if the gain and loss are worth the same amount; so they tend to overreact, assigning undue weight to the reasons for avoiding the associated action.[6] In common parlance, they throw the baby out with the bathwater. Because something feels wrong, they assume there is a logical reason to not do it, without an adequate examination of the risks involved. This can contribute to excessive risk aversion.

But emotional reasoning can also be a factor in overconfident exceptionalism. Entrepreneurs are particularly good at overriding concerns this way; in a sense, they have to be. *We feel good about this; therefore, we expect no problems* is an example of assuming that your emotions have more logical validity than they actually do.

Emotional reasoning often leads to self-fulfilling prophecies. For example, if your company is acquired, you may recall a similar experience from years ago. *This is just what happened before I was laid off.* Whether or not you are actually marked for dismissal, you feel the same mistrust, fear, and lack of commitment that you would feel if you were. Naturally, you are self-conscious, stiff, and resentful, thereby making leaders more likely to ask you to depart.

The flip side of emotional reasoning is rigid rationalism: *We came to this decision logically, so there will be no disagreement with it.* This is the misperception underlying the "economic rationalism fallacy"— the idea that a rationally defensible outcome will automatically lead to good feeling. *Everyone supports this downsizing because they have heard the rationale; they agree that it will make us a higher-performance company.* The layoffs may be necessary and justified, but they will not necessarily spark the emotions you think they will, any more than people are purely rational actors in economic situations.

The Power of Relabeling

This list of deceptive organizational messages is not an exhaustive catalog; undoubtedly, you will notice others. There is an infinite variety—as vast as humanity. Moreover, any deceptive messages— either brain or organizational messages—can interact and reinforce each other's power, even when they seem to contradict each other. People in a company can swing rapidly back and forth between unrealistic risk aversion and overconfident exceptionalism, with little or no awareness of this as a recurring pattern and with each new swing of the pendulum leading to greater exaggeration.

Relabeling gives you a way out of the trap. Remember that the source of the problem is not the message itself. There will always be deceptive messages, in everyone's brain and in every organization. You can't completely stop them, nor should that be your goal; indeed, to make that your goal would itself be a form of all-or-nothing thinking.

But you can choose how to handle these messages. You can help keep the organization from treating them as accurate. The first step is to raise collective awareness of them: *These messages don't represent us. They are simply things we tell ourselves, and the more clearly we see them, the more easily we can change.*

This act of relabeling may not seem like much in itself, but it is one of the most powerful things you can do as a leader. It establishes your voice, and your presence as a Wise Advocate for the enterprise, in a way that does not put you at risk: you're simply identifying a message that everyone is already aware of. To be sure, there may be powerful negative reactions. People may feel anger, worry, fear, grief, or stress when habitual thoughts, which people are attached to, are suddenly recognized as destructive forces. But you can put those feelings to use. Remind people that their anger and grief are often signs that a deceptive brain message is present, and that your company is starting to challenge it. Moreover, the act of listening to you and thinking about what you're saying tends to move people closer to the High Ground.

With relabeling, the absurdities and internal contradictions of deceptive organizational messages become more apparent. We know a company that went through an acquisition that took a year to complete. During that time, many staffers habituated an anonymous online message board, where gossip about the firm was rampant. During the negotiations, when it wasn't clear if the deal would go through, one common message on that board was: *An acquisition like this will destroy our firm.* Then rumors began to circulate that the deal was off. Now there were new messages: *We can't survive without this acquisition.* Only when people pointed out that these two contradictory messages couldn't both be true did the tone of the discussion grow more temperate.

Inquiry, not preachiness, is the key to effective relabeling. Don't say, "This message is wrong" or "Why do we even believe this?" Instead, engage in open-ended inquiry: "How did this message become part of our way of life? What problem were we trying to solve?" If no one has questioned the concept, the strategy, or the approach in years, simply asking questions like this will make it clear that these are not unchangeable precepts. They're ideas and practices that were adopted in the past and that can be reconsidered, once they are recognized for what they are.

For instance, suppose you are the leader of a fast-moving functional department that fields many urgent inquiries. You have come to rely on one person in particular for managing these inquiries. This individual—let's call him Dennis—has elderly parents who suddenly require care, and he is responsible for them.

Dennis asks you if he can change his work arrangements. He'd like to adjust his schedule to come to work early and leave early, so that he can manage his parents' care in the early evenings. Your first instinct is to agree without hesitation, but you find it difficult to send the email telling him yes. Something clearly doesn't sit right with you.

You realize you have several feelings at once. They're small feelings; this is not that big a deal. Yet they're significant enough that you've put off responding. Once you stop and reflect on it a bit, you realize that you are angry—angrier than you care to admit. And if you're angry, then *something must be wrong with Dennis for making you angry.* Underlying that feeling might be any of the four categories of deceptive messages that we identified:

1. Misperceptions of risk (excessive risk aversion): What happens when an urgency strikes and Dennis isn't around? You'll have to take care of it yourself. Moreover, this type of parental care can go on for years. You can't take the chance.
2. Misperceptions of value (all-or-nothing thinking): Dennis should not be giving you this problem to deal with. If he can't manage it on his own, there must be something wrong with him.
3. Misperceptions of proficiency (entrenched insecurity): You can't grant Dennis this request because then you'll have to do the same for anyone else who wants it. And you aren't that capable.
4. Misperceptions of validity (emotional reasoning): It doesn't feel right. Someone who spends so much time with his parents can't be trusted.

Now that you have recognized the messages behind the feelings, you see your anger and grief for what they are: emotions triggered by the messages themselves. You can see this more easily once you've shifted your attention away from the Low Ground. On the High Ground, you invoke different patterns of mental activity, more involved with the Executive Center. This makes more mental resources available to you as you figure out what to do.

The Ladder of Inference

One conceptual device helpful for relabeling is known as the Ladder of Inference. First developed by the linguist S. I. Hayakawa, adapted for organizational use by Chris Argyris, and then popularized by Peter Senge, it is designed for use at moments of great stress or fast-paced activity, when deceptive brain and organizational messages often hold sway. To relabel messages, you stop the conversation and then think through the situation with others. The Ladder of Inference gives you an easy way to do this: a series of cues for inviting you and other people in a conversation to relabel your deceptive brain messages together.

The ladder, shown in figure 4.1, is particularly valuable in teams or small groups. It is based on the idea that when you make explicit

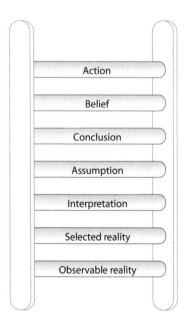

Figure 4.1 The Ladder of Inference.

the mental activity involved in shared assumptions and deceptive messages, it is easier to manage conflicts. It often takes the form of an exercise, in which two people who misunderstand each other compare views and resolve their disagreements. It helps articulate and understand the reasons why their perspectives have become so different.

At the bottom of the ladder is a rung we label "observable reality." (It has also been referred to as "observable data and experiences.")[7] These are the objective facts that would have been evident if the relevant conversations had all been recorded on video —the elements of the situation that both people (indeed, anyone, including an impartial spectator) would agree are clear and evident.

Then you talk your way up the ladder, coming to a mutual understanding of:

- The data you both selected: the facts and cues that you chose to pay attention to, based on your beliefs and prior experience
- The meanings you both added: your interpretation of the significance of the data you selected, which was already different from the other person's selection
- The assumptions you each made: the connections between your interpretation and your preexisting views of how the world works (which you may not have questioned or reflected on recently)
- The conclusions you each drew: your divergent points of view as to what is going on in this situation, based on your assumptions and interpreted reality, probably very different from each others' conclusions
- The beliefs you adopt: your unexamined perspective on what this situation "proves" and how it should affect your action—in other words, what the right thing is to do
- And finally, the actions you take based on those beliefs. You now make decisions that seem right to you because they are based on what you believe, but they seem incredibly wrong to others who have different beliefs about the same reality and facts

Note that only two rungs—observable reality and action—are visible to the group of people. Everything else occurs as a pattern of mental activity. The ladder itself is a series of labels for deceptive brain messages—a way of sorting out the sequence that leads you to self-destructive action.

For example, Lauren went through something very close to the ladder's hierarchy when she considered quitting the company:

Observable reality: Majid has gone to the founders of the company to "clarify" Lauren's directions.
Interpretation: He is doing this to undermine me.
Assumption: He doesn't want to take direction from a female, and he will not respect my decisions.
Conclusion: I have to fight for every decision I want to make.
Belief: Being a leader in this company means being in a constant battle with my colleagues, especially because I'm a woman.
Action: Lauren prepares to resign.

Though Hayakawa, Argyris, and Senge didn't use the terms *deceptive brain messages*, *relabeling*, or *Wise Advocate*, the ladder is deeply congruent with those concepts. A deceptive brain or organizational message is typically a leap up the ladder, drawing a conclusion that is based on interpreted facts but not really supported by the observable facts. If you return to the same messages, before long you are literally jumping to conclusions—missing facts and skipping steps in the reasoning process. These leaps often happen so rapidly that the conclusion seems inevitable and unquestionable. But with this guide before her, Lauren can lay out the whole chain of false logic. She can write out the assumptions ahead of time (calling on her Wise Advocate to help her) and then say something like this to Majid: "I don't think we've been fully clear with each other about our decisions on this new territory. Could I ask you to explain your thinking to me? And would you be willing to listen to mine?"

If she can make Majid feel welcome and reveal her own internal logic, then he will be in a much better position to reveal his own:

Observable reality: Lauren has asked Majid to postpone setting up the warehouse for expansion into a new territory.
Interpretation: Lauren didn't ask for my opinion, and that led to a shortsighted decision.
Assumption: When the founders told me to check in with them, they wanted me to keep an eye on Lauren.
Conclusion: The company leaders readily use us to keep tabs on each other, because they don't trust any of us (and especially not me).

Belief: Being a leader in this company means being in a constant battle with my colleagues, especially as someone from a different background.

Action: He goes back to headquarters to confirm (or "clarify") Lauren's decisions, without asking her first.

Neither Lauren nor Majid is eager to see the other fail. But they have gone so far "up the ladder," that they can no longer see past their mistrust of each other. Their interpretations, assumptions, conclusions, and beliefs have all been labeled "reality" in Lauren's and Majid's minds. Now they must relabel them—as just interpretations, assumptions, conclusions, and beliefs. That's the only way to stop jumping to conclusions. They have to bring to the surface the deceptive brain messages that have influenced both of them.

They do this by talking openly about their moves up the ladder of inference. In this conversation, they try to separate observable facts from deceptive brain messages:

What are the actual events that we both agree took place?
How did we each come to our conclusions?
What do we believe now, having considered the events together?
How might we test our new beliefs?

One such test was the conversation itself. Lauren realized that her deceptive brain messages (*He is trying to undermine me because he doesn't want to work with women*) were different from the reality (*He is confused about what his role should be, and I could make it easier*). She asked him about this, and they talked more openly about how each had assumed the role of outsider, not realizing the other felt the same way.

Another test occurred when Majid went back to the founders and asked them what they thought his role should be. It wasn't until then that he was sure about the gap between his deceptive brain messages (*Even though they don't trust me, they want me to keep tabs on her*) and their actual interests (*They want us to work together to build the business*).

We don't suggest that the Ladder of Inference, or any technique, solves everything. But any technique that abets relabeling—that helps you consult your Wise Advocate and identify your deceptive brain messages as cognitive distortions, not as perceptions of reality—will be helpful and enhance mindfulness. Indeed, Lauren and Majid continue to misunderstand each other from time to time. But they have opened the door to a new habit: talking together when there is a misunderstanding and trying to come to a common understanding of the source—the deceptive brain messages and corporate cognitive distortions underlying it.

The Leader Who Relabels

Your ability to act as a catalyst for the larger system around you—to play the role of Wise Advocate for your organization—starts with your ability to relabel. Consider, for example, zero-based budgeting—the act of looking at a department's budget in terms of how it contributes to the company's strategy, rather than basing it on what was spent last year. The old budgeting approach may have produced a feeling that people crave: stability and abundance in the face of uncertainty and competition.

Now imagine that it's halfway through the fiscal year and the leadership of the company says something like this: *Our old budgeting behavior, taking the previous year's figure and adding 3 percent, is just one approach. We're going to consider alternatives for next year.*

Chances are, of course, that there is a legitimate cost-management issue involved. The company may be overspending. But people feel threatened because the "3-percent-plus" approach to budgeting feels good; it provides security and validation in the face of uncertainty. They crave that feeling, and the new approach, whatever it brings, has already diminished it by raising uncertainty. The resulting resistance could be felt as "fight or flight" panic: *If they cut our budget, or force us to lay people off, I don't know how we'll manage.*

When that happens, relabel the resistance. Look closely at the source of each deceptive message circling around the enterprise. *What kinds of messages are being sent by top leadership? What kinds of messages are emerging from other parts of the hierarchy? What are the messages on both sides seeking to have people do? Will those actions move you in a better direction?* By looking at the messages dispassionately this way, you trigger the Executive Center and the Wise Advocate circuit, even when your sense of resistance is prompting you to stop.

The point is not to discount messages because they evoke strong feelings, but to recognize the role that feelings have, over the years, in the habitual nature of the messages. Instead of criticizing the messages, articulate what you think they're trying to say. The more context you can create for seeing organizational deceptive messages clearly, the more effective you can be at helping others move their attention to the High Ground.

Organizational systems theorist David Kantor, author of *Reading the Room*, calls this bystanding: providing a perspective on the action taking place in a meeting in a way that helps to reconcile conflicts or problems. When this is done deliberately, says Kantor, it is transformative for the entire group. His description of the practice will remind you of the Wise Advocate: "Rather than [acting] on the basis of your own subconscious behavioral hot buttons, first learn to step back mentally and extract yourself from the action in order to see what is occurring."[8] Have you seen this pattern before? Can you give voice to the messages in the conversation? If so, then you are relabeling the conversation for everyone, in the same way that, as an individual, you would be relabeling the mental activity in your mind.

There are many examples of leaders who have done this, but one of the most compelling stories is that of Ian McRae, the former CEO of Eskom, which is the largest power utility company in South Africa, responsible for 95 percent of the electricity in the country. Eskom was one of the first companies to move away from apartheid, the authoritarian system of racial segregation that governed South Africa from 1946 to 1994, effectively excluding nonwhites (who make up 79 percent of South Africa's 47 million people) from the nation's economy and politics. In the late 1970s and early 1980s, Eskom was legally prohibited from providing electricity to black communities. Most of the country's black citizens were illiterate (government policies made it difficult for them to be educated), and there were almost no black professionals in the country. McRae, who would not become CEO until 1985, was then in charge of power stations. He began to realize that the company's apartheid-based policies were unsustainable. There were, literally, not enough white workers with the skills needed to staff those plants.[9]

To begin training black workers to fill these positions, McRae had to overcome a number of powerful deceptive brain messages and corporate cognitive distortions. For example, business managers and employees often believed that *black employees are only fit to be unskilled laborers.* This false assumption was deemed plausible because many black people in South Africa were barely literate; apartheid had restricted their education. "If [black employees] were to fill skilled jobs," McRae later recalled, "they would have to become literate. We would have to train them to read thermometers and pressure gauges, and to keep written records of the equipment they were operating."[10] In decisions like these, McRae started reframing corporate cognitive distortions. *Black employees can be absolutely capable of filling skilled roles.*

Next he began setting up meetings at each power station with the various constituents: trade union representatives, plant managers, and black laborers. He did not present the new policy as a fait accompli, nor did he insist on his point of view. Instead, he asked people to talk about their assumptions, then subtly but persistently relabeled and pointed out that these were just assumptions. After all, there were counterexamples: black people who had successfully become professionals in other countries, and in South Africa as well.

"I went around asking people what they thought of [our proposed shift to a multiracial workforce]," wrote McRae, "presenting my view, and listening to theirs. Most of them did not agree with me, but I kept asking for their concerns. One of the most conservative power station managers phoned me at home. 'You keep me awake at night,' he said, 'and I'm starting to think you're right.' Now he accompanied me when I spoke to Eskom managers and engineers. What better consort could I have, in that political environment, than a right-wing leader? I also had to talk to the Black staff members, who said, 'What if we fail at these new opportunities? What will happen to us then?'"[11]

Because of the time he took raising awareness of these deceptive brain messages, there was relatively little resistance when, in the early 1980s, McRae started introducing black South Africans into the ranks in the new position of "operating assistant" and providing them with the training to develop their technical skills. He removed the existing educational barriers so that nonwhite operators could move into the position of shift supervisor. A corporate cognitive distortion had been challenged and changed.

We saw a similar example of relabeling in chapter 1, when Alcoa's Bill O'Rourke insisted that the conventional wisdom about Russian business—that bribes were an inevitable price of doing business—was not necessarily reality. It was a self-reinforcing perception. We can also point to the story of Aetna CEO Mark Bertolini, which we told in chapter 3. Bertolini, as you may recall, established an across-the-board raise for the lowest-paid employees at Aetna, giving them an average pay increase of 11 percent to $16 per hour. When he first proposed the move, his human resources staff replied that the costs would be too high: it could cost as much as $10.5 million per year. He countered that, while the estimate might be right, the assessment of the costs was just an assessment, and it depended on the business impact. In so doing he was exercising discernment, a key Wise Advocate function.

"Our accepted figure for turnover costs was $27 million per year, but that was only voluntary turnover," Bertolini later recalled. "I asked for total turnover costs. How many people leave involuntarily? How much does it cost to hire their replacements? How long does it take to train the new recruits? We looked at absenteeism, rework, productivity, dissatisfied employees, and our net promoter scores [a measure of survey respondents' enthusiasm] in recruiting new employees. . . . We figured out that our total turnover costs were $120 million per year. By that measure, $10.5 million looked like a low-risk investment."[12] In other words, the idea that the costs were too high was a truly deceptive organizational message—a corporate cognitive distortion so prevalent and pernicious that it took a major research effort to relabel it.

It may seem as though there's a noble component to the stands people take against deceptive brain messages, and it may sometimes feel that way, but many times the corporate cognitive distortions have to do with simple business realities. Nor do you have to be a senior executive to relabel a deceptive organizational message of that kind. Here is a story from an oil company, told by the operations foreman at a refinery:

We were going to shut down a dehydration facility that was pumping 1,500 barrels a day. I had an operator who refused to go along. "You're doing it all wrong," she said. "I can keep you online and running." She had a bunch of us red-faced and angry at her, but she refused to budge. We showed her diagrams from the engineer. But she stood her ground.

[Earlier in my career,] I'd have said, "The engineer says this is what you do because he's an engineer and you're an operator." Instead, I let her hold up the project a few days to solve the problem, and six to eight weeks to implement her solution. There would be times when I would go out there and she would be sitting on a brick cement wall, motionless. . . . I had to force myself [to allow her to continue.]

In the end, I give her all the accolades. She came up with a process where I never lost a barrel of oil! What she did was very, very intelligent—a simple procedure that let us bypass the treater that was causing the problem. She determined that she was not shutting down.[13]

The deceptive organizational message here was simply that *the knowledge of engineers is worth more than that of operators*— a misperception of value. The relabeling took place without a word.

Relabeling at an organizational level will yield many benefits. It gives a name to the tension and threats that people feel together. At the individual level, it helps calm the Warning Center. At the organizational level, it can stop the debilitating and demoralizing flows of gossip that often occur through back channels and corridors. It can equip people emotionally and cognitively to deal with a broader range of challenges than they have had to deal with before, because it brings members of the group naturally to the High Ground. People are thinking about their thinking, approaching a state of metacognition. Once they have moved in that direction, they are ready for the next step: reframing.

CHAPTER FIVE

Reframing Your Situation

THE FINANCIAL CRISIS of 2008 highlighted several widespread deceptive organizational messages that had quietly misled the banking and investment sectors for years. Before the crisis, for example, it was easy to believe that rapid growth would continue indefinitely. *This is not a bubble; we've finally learned how to sustain financial success.* Or that private equity fund managers knew best how to turn around a company; after all, they put themselves on the line every time they took a controlling interest. They were smarter, more decisive, and immune to the entrenched loyalties that build up in any organization.

"They charged high fees," recalled Marcus Simpson, the head of the private equity program for the Queensland Investment Company (QIC), "but they made money for their clients, including us. They got very rich. And our staff tended to think of them as masters of the universe."[1]

QIC was owned by the provincial government of Queensland, Australia (the northeastern province, whose capital is Brisbane). As a major sovereign wealth fund, and a relatively passive investor, QIC let third-party private equity funds manage most of its investments, delegating many decisions to them. These intermediaries charged high fees, but their investments had always paid off. Investing through intermediaries was also a good stance politically; it kept QIC relatively free of conflicts of interest.

As you might expect, the managers of the private equity funds agreed that they were masters of the universe. One of them told Simpson, "You guys are doing well now, but you're going to screw up, because only we can hire and pay up for the best people." Behind statements like this, there were other deceptive messages about the nature of QIC's business: *Entrenched management is inherently weak because its people don't make tough but necessary decisions. Complacent, bureaucratic companies need to be cut back to the bone before they make money.*

Another deceptive organizational message was grounded in outward-looking optimism: *The economy is going to grow indefinitely. The experts have figured out how to avoid downturns.* As we noted in chapter 4, economists Carmen Reinhart and Kenneth Rogoff argue that this deceptive message is present in the lead-up to most financial crises.[2] Like many other investment firms in the early to mid-2000s, QIC had based its financial projections in part on this assumption.

Then came the economic crisis. By 2009, a number of funds were faltering. This was a wake-up call for QIC. Returns slipped, and QIC's leaders had no choice but to relabel—to recognize that their habitual deceptive messages were only messages, were probably false, and could threaten their existence. *This time is not different. Instead of enjoying easy, continued growth, we have to dig ourselves out of a hole.*

In cases like this, relabeling will only go so far. You have to reframe—to replace those messages with different narratives, composing a new story that will be just as compelling but better for you and for the organization. This new message is one that can help you play the role of Wise Advocate; it stands for the long-term value and purpose of the enterprise.

Marcus Simpson, who had joined QIC in 2005 from a London-based private equity firm called Altius Associates, had already been skeptical of the old narrative. The financial crisis made reframing it more urgent. Meanwhile, the outside fund managers were also facing existential threats. "With the declining global supply of fund capital," Simpson later recalled, "[the funds] had less capital to manage under their juicy fee structure, but they still wanted to invest." They began to press QIC for more capital. To get out from under this pressure and maintain its requisite returns, QIC would have to do its own investing."

Simpson began to spend more time in mindfulness practice; he also relabeled his views of other people in the company on the team, which triggered a fair amount of mentalizing. *What are my colleagues' assumptions about investment? How does that affect what they will do next?* He found that his team members recognized the reasons why QIC should do its own investing directly, but they lacked confidence. "Some members of my team would say, '*Those fund managers are amazing. We're never going to have that skill set.*'"

This could only be addressed by a reframed message: *We have to manage much more of our own investment. We are (or can be) capable of doing it as well as, or even better than, the external fund managers. And if we don't succeed at first, we will figure out the solutions.* Backing this up was the recognition that the value of a deal depends on a nuanced and varied group of factors, including the ability of diverse players to think and work together. *We are culturally well-equipped to make these kinds of deals.*

Simpson and his colleagues gave the name "modern private equity" to their new story. It was an alternative to investment by external activists. "We went back to first principles of what drove superior returns in private companies," Simpson later recalled. "We wondered if we could develop more control over what we did, investing directly in companies and partnering with their managers, with a better mix of risk and return than pure fund portfolio could give us." This was, in effect, a different theory of successful private equity: Instead of taking costs out rapidly and flipping the companies, QIC would foster relationships with the existing management of the companies in which they bought a stake, help them transform their operations, and gain results from collaborative learning between the investors and the enterprise.

Simpson believed that if they could make collaborative investment work, reframing the act of investment as investors and managers learning from each other, it would give them a confident approach that would stand them in good stead for years. Indeed, they could surpass the track record of the "masters" at the private equity funds—especially in regions like China, where few investors had experience. "We invest outside conventional areas (for example, in agribusiness)

and take much more control over the portfolio than traditional investors would," says Simpson. "We get more involved and we incorporate
new learning, new findings, and new developments as we continue to
refine this modern model."

A reframe like this may sound simple from the outside, but it
involved exercising the Wise Advocate function on the part of many
of the staff. They had to change their attitudes and venture into the
unknown. They were forced to adopt new investment-management
skills, to be more authoritative about new investment areas such as
Chinese ventures, and to be more assertive—with themselves as well
as others—about the distinctive value they brought as investors.
One factor that helped was the arrival of a new CEO, Damian Frawley,
in 2012. "When I joined QIC," Frawley recalled, "I found some of
the most entrepreneurial and commercially minded fund managers
I'd ever worked with. But because they sat within a government-
owned business, neither they nor anyone else was fully aware of this
skill set."

QIC's new "modern private equity" narrative had to be complete
and comprehensive enough to withstand resistance and cover any
concerns people raised. "We had to transform the thinking of our private equity team making the investments," says Simpson. "We also had
to get the support of our chief investment officer, our investment com-
mittee, and the board. We had to give them confidence that we could
accomplish our goals." Adds Frawley, "We had to show them, particularly the fund managers, that there was a material benefit in signing on
to a more unified and slightly more interventionist approach."

The reframed way of doing business turned out to be highly
successful. QIC has produced consistently outstanding returns:
19.1 percent per year on average since 2005, which is more than
5 percent above the public market equivalent.

As they developed these skills, Simpson became a consistent source
of guidance and support for others at the enterprise. Repeatedly, he
reminded QIC staff about their ability to take more control over the
portfolio than traditional investors would. "We get more involved,"
he would say, to both insiders and outsiders. "We incorporate new
learning, new findings, and new developments as we continue to
refine this modern model." In effect, he took on the role of Wise
Advocate for the company, projecting his own internal attitude shift
onto the company at large.

The net effect was a deepening of judgment within the enterprise—which made QIC not only more successful than its former private equity contractors, but also more consistent in its approach. "There's both art and science in what we do," Simpson later recalled. "On the art side, we gain a fair amount of skill by watching human behavior, our own and that of the people we invest with. For example, after the financial crisis we were constantly asked, 'Why aren't you doing more in Asia?' We were looking at the behaviors—at the way people acted with respect to pricing equities. It wasn't the financials that convinced us to be cautious; it was the quiet recklessness, compared with the way people had acted before. Quite often, we said, 'That's not for us,' on a deal that others leaped to make. And we prospered accordingly."

Reframing Is Step Two

Reframing is an applied use of a primary Wise Advocate function: the act of building a new constructive narrative. It recasts the prevailing messages of your brain, and the enterprise you are part of, with a new logic. This makes it easier for everyone to pay attention in more constructive ways. With this new narrative, you have a new source of internal messages—both brain and organizational messages—and the habits of the past begin to change.

In this step, you are as deliberate as you can be in choosing the new messages to adopt. In relabeling, you said, in effect, *These messages are not who I am.* Now, in reframing, you answer the question "What messages, then, reflect the me that is really me?" or, organizationally, "What messages reflect the organization that is truly ourselves?" Your new narrative is aspirational: it doesn't focus on what you fear or want to change away from, but on who you want to be, what you can accomplish, and what your organization can become.

Developing this new mind-set is a complex, iterative endeavor. It doesn't happen all at once. It is both a dramatic leap away from the familiar and a narrative based on who you already are. You can think of it as changing the standards by which you choose what is important to you, deliberately using self-directed neuroplasticity to accomplish this. In chapter 2, we compared neuroplasticity to a search engine; it is always easier to search for things you already pay attention to.

Now you are deliberately paying attention to a different pattern. It is a bit like reprogramming the search algorithms; you don't want to be limited by listings that fit with your past behavior, but you don't want to leave behind your past altogether. You want a new story line that is stable enough to hold onto, but different enough to allow you to break free of past limits, and more in tune with reality than with your habits of mind.

You know you are reframing successfully when levels of anxiety and mistrust in your organization start to go down. That's because the new narrative is conscious of what your organization realistically can be, and of the aspirations of everyone within it. For us, this is linked to the concept of the "true self." As you may recall from chapter 1, the true self is a core of personal belief and identity, the person you are and aspire to be, the person who embodies the values and goals at your core, developed in consultation with your Wise Advocate. A focus on the true self is particularly apt for the many individuals who are stuck at work, afraid they are inadequate for their role or unwilling to take the chances they need to advance their careers. When they hear someone give voice to aspirations like these, they can transcend the impediments that keep them feeling stuck.

Since every circumstance is different, we can't describe common reframing narratives. They aren't like deceptive brain or organic messages, with archetypal recurring patterns. But we can tell a few stories of successful efforts to reframe—at the individual and organizational level—and then describe the factors that make a reframed narrative effective.

Work-Life Balance on the High Ground

A good place to see reframing in action is in the common problem called "work-life balance." Many professionals simply don't have the time to do justice to the varied pressures on their lives. They have demanding jobs that they have worked desperately hard to achieve and that still don't pay them enough to "throw money" at every problem that comes their way. They may have children, which in industrialized countries means they will spend hundreds of thousands of dollars on care and education before their children are independent. They have their own health and well-being to take care of, and they may also have relatives (such as aging parents) to care for. They have commutes, communities, friends, and private ambitions that all demand time and attention.

Indeed, as *New Yorker* writer Adam Gopnik points out, this problem is probably directly related to industrialization. In literature before 1840, there are few if any references to being too busy; by the early twentieth century, characters in books are continually short of time. What happened in between? The telegram, the railroad, the telephone, and the automobile brought people (at least those in cities) in touch with many more people. If you want good work-life balance, you have to know fewer people.[3]

We know that's not likely. And many business leaders have the income and resources they need to manage this problem—somewhat. But as they get older and better established, they don't find themselves able to slow down. Instead, the pressures to achieve, compete, and contribute tend to increase. The deceptive messages that they continually hear from their own brains—about the difficulty of the problems they face, the fierceness of the competition, and their own shortcomings—add to the anxiety they feel and make it all the more difficult to cope.

No wonder executive coaching sessions often focus on time pressures and work-life balance. The problem persists because there are no easy answers; managing the time and energy available to an individual is a lifelong discipline that each individual must handle according to his or her priorities. Reframing this problem, and bringing the Wise Advocate to bear on it, is the only sustainable solution that we know. The internal gyroscope of the Wise Advocate's guidance can help you continually fine-tune the delicate balance required.

Josh leads a functional division at a large consumer products company. Like so many middle managers, he has faced the perennial problem of feeling overworked. For years, he spent his days ticking off items on a massive to-do list and was in the habit of blowing up at his wife and children when they asked for his attention. His typical response was "I'm busy." At work, the more he accomplished, the more he was given to do, and he never felt he could say no to anything, particularly when the company's revenue growth slowed down and everyone was expected to help solve the problem.

Emily, a university administrator we know, had a similar plight. She had twenty direct reports and managed a large student service unit.

She was continually concerned about not having time to manage family commitments. "I felt like I couldn't adequately control my work-life balance," she said later. "Emotionally, I felt that there was nothing I could do about it. I thought of time pressure as something being done to me."

If you are like Josh and Emily, this issue probably seems insoluble—at any deeply satisfying level. You are always disappointing someone, and if you're satisfying everyone else, then you're probably disappointing yourself by not taking the time you need for yourself. John and Emily knew they were not alone; they knew how many people wrestle with this problem. But that didn't make them feel any less stuck.

Both of them came to a solution through coaching, which led them to relabeling and reframing. Relabeling, as we saw in chapter 4, involves learning to recognize the deceptive brain messages that contribute to any problem. For anyone struggling with work-life balance, the deceptive messages probably involve misperceptions of value: *I have to solve every problem right away, or I won't be worthy. I cannot say no. I will never get the time, resources, or support I need, so I have to make the best of it. Every priority is equally important. My work is more important than my personal life or my health—in fact, it is more important than anything.* Relabeling these thoughts—seeing them as thoughts that are not completely true—is a critical step.

But you must also reframe the situation. Josh began with a recognition of something he often lost track of: his love for both his work and his family. When he caught himself about to say "I'm busy," he stopped and reflected silently to himself on a new, deliberately chosen message: *I have a full and abundant life.* The phrase reminded him that his basic problem was not being busy, or stress. What he hated was the feeling of having to make trade-offs between his work, his family, and himself. But that frustration was simply one more message from his brain.

"I sorely want to do everything, including take time for myself," he said to one of us. "How can I pay more attention, within my full and abundant life, to all the priorities I have for myself, without feeling like I'm dropping the ball, and without losing my true self in the process?"

Reflecting on that question brought his Wise Advocate into play and directed his mind to the High Ground. One new message that came to him was *I can and will take the time to think through complex decisions.* Before giving the "automatic yes" response that was deeply ingrained in his Habit Center for many years, he now takes a mindful breath to collect his thoughts and responds with "Let me come back to you within twenty-four hours with an idea for moving forward." This gives him the time to think it through—to make a clear assessment of what is possible, realistic, and manageable—before he commits to a deadline (or not).

Josh still has an aversion to saying "no." He knows that in his company, a plain "no" will often trigger a threat response—a disproportionately emotional reaction—in himself and in the other person he is talking to. But that doesn't mean he just says "yes." Borrowing a phrase from management writer Jesse Sostrin, he has reoriented himself to always say "yes, if . . ."—to set conditions on his own terms. "Yes, I can deliver that outcome if you will provide me the first half of the project complete," or "Yes, I can arrange that if we can line up the right staff people to execute."[4]

When he goes home and his family members ask for his attention, Josh now reframes and says, "I've had a full day helping a lot of people, and I need to rest a little while before I sit down with you and give you the attention that you deserve." At work, he has found more effective ways to reframe and say, "This is a full body of work we're doing, and we need to set better priorities." He sets aside time for important personal goals, such as exercise and family. He also spends at least half an hour per day in quiet contemplation, which often involves a breath awareness exercise like the one described in chapter 3. This was difficult to arrange at first, but it has now become habitual. He is accustomed to blocking out time with no phone calls or email. He spends as little time as possible on the "nice-to-do" tasks and explicitly communicates to relevant stakeholders why these tasks will not be met.

Emily also reframed and replaced her deceptive brain messages about time pressure with a new narrative. *I have my position because I am powerful*, she said to herself. Not only that, she added, but *I have a formidable talent for organizing systems*. She had never applied that talent to her work-life balance. She knew that the first step in systems change is to establish a clear parameter of success, so she set one: *I will no longer work more than fifty hours per week, but I will make the most of those fifty hours*. She didn't announce the new policy until she had succeeded in executing it for two months. By then it was permanent, and it gave permission to her staff to do the same. This was the beginning of a two-year-long maturation in which her team members bolstered their productivity and enthusiasm, as she was bolstering her own.

Reframing your messages around time and energy may lead you to organize your life differently. You may end up scheduling your time in blocks—allocating a set number of hours per day to family and self and ruthlessly sticking to that schedule. You might set clearer priorities for your role and stick to them, delegate more, reorganize your team, remove some of the unnecessary practices and checks that take up time in your enterprise, or all of the above. What matters most, of course, is that these new practices bring you in touch with your Wise Advocate and take you closer to your true self.

The more capable you become at this, the more capable you will need to be. Others will demand your time more when they see your productivity and value increase. In fact, this has happened to Josh. He sets aside several periods a year when, it is agreed in advance, he is unavailable—in part so he can regenerate, and in part so that he can spend a week training others to assume more of his tasks and accomplish them in similarly brilliant (or even better) ways.

Of course, this requires self-discipline, but here you are helped by the executive function that you strengthen as you cultivate your Wise Advocate and move onto the High Ground. Reframing deceptive brain messages—including the idea that you can do everything demanded of you—becomes a continual practice that takes you away from the transactional people-pleasing nature of expedience and closer to your true self.

Reframing in Organizations

Many people in leadership positions must take their enterprises or organizations through a significant change. The ability to reframe effectively is an essential component of this.

One individual who understands the value of reframing is Keith Bailey, currently a strategic adviser with the turnaround firm Amplifi Governance. Bailey has spent much of his career as an interim CEO, joining companies in trouble for a year or two and helping them emerge from underperformance to growth. "In many cases," he says, "I have to educate the board and the management team, which means providing clarity and support—giving them the perspective they need to develop or approve a new strategy with confidence." He plays the role of Wise Advocate at the most senior levels of companies, and sets them on a track to become Wise Advocates for themselves.

One such company was TransPacific Industries (TPI), a large recycling, waste-management, and industrial-services company based in Brisbane, Australia—and, at one time, the largest Australian waste-management enterprise. In the early 2000s, TPI was on a roll. Its founder, Terry Peabody, had built it up from a small coal-ash-recycling company, making about fifty debt-fueled acquisitions along the way, and had become a billionaire in the process. He was a press-shy business leader, nicknamed the "Golden Garbo" for his company's rapid growth, and known for his idiosyncratic strategies. TPI's expansion culminated in 2007 with the purchase of a waste-management business called Cleanaway for A$1.25 billion (about US$1.1 billion)—an extraordinary amount for that industry in that region.

During the 2008 financial crisis, the bottom fell out of the industrial-waste-removal business, one of TransPacific's most vital sources of profit. TPI's share price, which had been A$9.96 (US$8.96) in mid-2007, fell below A$1 (US$.90), and it received an A$800 million (US$648 million) bailout from the private equity fund Warburg Pincus. After Peabody retired in 2010, the company went through three chief executives in rapid succession. It fell so close to bankruptcy that in 2011, multiple turnaround and restructuring activities were running concurrently.

One troubled unit within the company's complex structure was a manufacturing division that had been assembled from nine separate acquisitions. In 2011, the board hired Keith Bailey to help pick up the pieces. Because it was assumed that this division would be broken up or sold at a loss, Bailey was placed in an interim general management role. "There was no position description," he later recalled, "just: 'Here are the keys. You're on your own.' My directive was: find out the problems, fix it quick, get the business back to profitability and position it for divestment within nine months."[5]

The assignment represented an enormous challenge. TPI had "so much debt it almost fell over," according to one journalist. Its acquisitions had left it with many internal groups whose leaders barely communicated and often worked at cross-purposes. TPI had landfills that were running out of room, overdue compliance costs related to new environmental regulations, and a reputation for operational waste and inconsistency.

But the biggest problem the company had, arguably, was the stories it told itself—deceptive organizational messages embedded in its culture and repeated throughout the company as a matter of course. During the high-flying years, the messages had been grandiose misperceptions of risk: *We're extraordinary. We can manage anything that comes our way.* Now they were bleak: *No one can save this company. There's going to be a bloodbath. It's everyone's fault but mine.*

None of those messages turned out to be accurate. But they would have dominated decision making at TransPacific's C-suite and business units, and led it to further decline, if Bailey and his counterparts in the rest of TPI had not addressed them directly. During his two years as general manager, Bailey held repeated meetings in which he brought these covert assumptions to the surface. In effect, he shone a light on the invisible ruminations of the firm's culture, raising open questions about the pervasive, collective mental habits that people took for granted.

This had a powerful effect, because people hated those messages, even when they didn't agree about anything else. The manufacturing division was back to breakeven after nine months and profitable within a year. It was sold whole as a going concern at the eighteenth month. This preserved shareholder wealth and most of the jobs in the manufacturing division. The TransPacific Industries business has had its ups and downs since then, but it continues to operate today under the name of a former acquisition, Cleanaway.

Bailey's turnaround of TransPacific International went beyond raising awareness of existing deceptive messages. It also involved reframing—replacing the old messages with a new conception of the company's potential value.

Late in 2011, on the second day of his assignment, Bailey brought together the top fourteen managers of his division—seven in a conference room and the other seven dialing in from far-flung cities. He summarized his thinking on a mind-map document—a single schematic page laying out all elements of the turnaround strategy in graphic form, showing how they related to each other. The map represented a new narrative: *Yes, these problems are serious, but we are capable of fixing them ourselves, if we overcome our internal difficulties and change our practices.*

The map was simple and clear enough to be shared with everyone in the company, from the senior-most levels to the factory staff. It set the context for the painful measures the company would have to take during the next two years. These included huge cuts in staff and other costs, to "stop the bleeding." They would also need major operational changes, applying the lean thinking approach that had helped many companies instill quality practices at lower cost. Finally, they would divest some major underutilized assets in a way that allowed those assets to survive—and maybe do better—in other companies that were better suited to manage them.

Though the holidays were approaching, Bailey insisted they start executing the new approach right away. "I explained the urgency of change," he recalled. "I had three weeks to do site audits, meet all the key managers and conduct the high level assessment before most sites closed for three weeks. I could not wait until they came back in January." When he asked for their reactions and input, most people were skeptical, and he understood why. "I had not met any of these managers before. I was new. There was low trust and high personal stress. Corporate management could shut down the business any day."

During the next two months, January and February 2012, Bailey followed up with a series of in-depth meetings and training sessions, at least one in each of the company's eight main office locations. He reassured people that the pervasive rumors of a bloodbath were not true. He also said he understood that their ability to come up with new ideas had been impaired by the emotional impact of the past year, including the natural drive for self-preservation, which had led many of them to blame others.

For one pivotal training session with about thirty functional leaders and key managers from all eight sites, he asked participants to prepare in advance short talks about what they personally wanted their businesses to achieve. After the discussion ended, he passed around a statement of his own position and objectives, which he had typed up the day before. "It turned out to be very aligned to what they had just said. It blew them away."

With that action, he reframed their sense of each other. They had all believed the deceptive organizational message that they were in competition. Now there was a new set of messages to get across: *We're in this together.*

"We then went into a formal review process. I asked them to identify the waste and key failure points within their processes that prevented products being delivered profitably and on-time, and how they and their teams could address them. I kept my focus on the overall objectives, and avoided any blame mindset from emerging—which can easily happen when people are in distress. Everyone contributed." Within five hours, they had identified more than 130 problems and agreed on the top seven that needed to be fixed first. They also agreed on who would drive the corrective actions, with a support team nominated for each. These were all broad common issues that had affected most or all of the eight sites.

This reframing—*we know what to do and we are prepared to do it*—brought the group from ambiguity and conflict to a sense of shared purpose. Instead of other sites being seen as part of the problem, they collectively realized they were part of the solution.

Another company that benefited dramatically from reframing was Ameriprise Financial: a US$7 billion company, spun off from American Express, that is the leading source of financial advice in the United States. This time, the reframing occurred before the 2007 financial crisis. In early 2007, some of the Ameriprise leaders noticed that they were not doing enough to help their clients. Typically, when stocks fall sharply, a rational investor should step back and wait for a signal of what is going to happen next. But many investors rush to sell instead. It's natural to feel like selling: stocks are falling, and nobody wants to lose. But when investors sell during a price drop, this exacerbates their losses, because stocks often rise again soon afterward. The deceptive organizational message could be summed up this way: *Buy when you feel good about a stock. Sell it when you feel scared.*

Ted Truscott, CEO of U.S. asset management at Ameriprise, stated the problem with this deceptive organizational message succinctly: "Remember, when you have the price, you don't have the proof, and when you have the proof, you don't have the price."[6] In other words, by the time investors felt comfortable with a stock (the "proof"), it was probably already priced too high to be a good investment. By seeking reassurance, investors were undermining their own portfolios.

Ameriprise counselors were supposed to know better, and many of them did. But they, too, fell prey to the same emotionally driven habits. "Being a great financial planner and advisor requires not only technical expertise," concluded Kris Petersen, then the Ameriprise senior vice president of financial planning, "but an understanding of how people make decisions. Our clients are misbehaving with their money, and we have to do a better job of helping them." In other words, even though the advisers knew what to do individually, the Ameriprise organization was not giving them the support they needed to resist deceptive organizational messages. They knew better intellectually, but they had no alternative narrative to offer instead.

When the leaders realized the problem, and how they were misguiding their clients as a result, they took the opportunity to reframe the narrative. This could have been a moment of retrenchment—a mutual effort to deny the problem and give people the same deceptive advice they had been offered in the past. Instead, the Ameriprise leaders embraced this new message: *We take seriously the maxim buy low and sell high, even when it makes people feel uncomfortable.*

The reframing effort involved a great deal of conversation. The Ameriprise leaders reminded each other of the pride they all felt about building a better future for their customers, and they talked about the study results and the suggestion that they should enhance their own practices. Out of this conversation they developed a new approach to their guidance, one that offered a much higher level of success for their clients but was counterintuitive to everyone—including themselves. It repudiated the feeling of safety that went with following the trends. But that would be OK.

Jeff Marshall, a franchise leader in the Pacific Northwest, moved rapidly to put in place a new training program to incorporate this approach into the company's practices. At first, response was limited. Even many who were initially enthusiastic expressed doubt when they discovered that the training would take several months. But interest grew after the financial crisis began. From now on, they would be the firm that offered counterintuitive advice, the kind of advice that allowed investors to win: *Buy stocks when they are undervalued, before they have risen enough to be popular. Don't wait for the proof.*

Of course, doing this would require Ameriprise to conduct a much more attentive form of research. They would also have to hone their collective judgment—the ability to select potential winners early in their rise. Overall, they have done better ever since.

These two examples, and the example of QIS at the beginning of this chapter, show the link between reframing and business success. Any conventional theory of business success will ultimately be wrong, because business is continually changing in response to competitive realities. Thus, yesterday's success is today's constraint on success. Every successful strategy must involve reframing—an explicit move away from the deceptive brain messages and corporate cognitive distortions that represent the successes of the past but systematically understate the potential failures of today.

Reframing and the Placebo Effect

Reframing efforts, like innovations, don't always succeed. And even those that succeed in the short run don't always last. In this respect, reframing is like any change initiative: the greatest problem is that the impact wears off. Iliyan Ivanov, whose psychiatric research at Mt. Sinai Hospital focuses on addiction and adolescent psychiatry, has studied the factors that motivate behavior change. Emotionally charged events—including threats to one's own life—are not powerful enough as motivators. "When a person who smokes finds out their friend passed away of lung cancer," says Ivanov, "they probably would stop for a week or two. But as you distance yourself from the emotional impact, you don't make the same investment over the longer term."[7]

The key is the difference between expectations and commitment. Reframing asks you to make a commitment to a new idea, and you undoubtedly desire to make the change. You will also have expectations about whether or not the change will work. In many cases, your commitment will feel strong, but you will also protect yourself by holding low expectations of success. You will be fully prepared, in your mind, for the fact that it may not work.

It might seem that this will help you move forward. Surely low expectations must be realistic; that way, you are guarded against disappointment. And when you make progress, the surprising results will give you momentum. But studies have consistently shown that the opposite is true. High expectations lead to better results, in part because they are correlated with strong effort.[8]

One source of knowledge about expectations is research on placebos—medications with no medicinal value given to subjects in experiments. The placebo response is a consistent finding: those who believe they have been given medicine will experience relief of symptoms (particularly pain), whether or not the belief is accurate.

Donald Price, a distinguished researcher at the University of Florida, studied the placebo response for many years before he passed away in 2016. He found that the expectation of a positive result—for example, expecting pain relief when an inert cream is applied—will make a difference to the actual experience of pain. Price set up a carefully executed series of experiments with volunteers who had a medical condition that made them particularly sensitive to certain kinds of pain. He gave some subjects a placebo along with a specific suggestion that led them to expect a reasonable chance of pain relief. He told them they would receive a drug known to cause pain relief in most people. Actually, the drug was saline. This expectation, in itself, was enough to relieve pain as effectively as real medicine would. It also calmed down the brain's pain and visceral centers—the thalamus and insula (a key part of the Warning Center). During and after the procedure, the volunteers reported experiencing almost no pain—as if they had taken novocaine.[9]

For neuroscientists, this is a fascinating finding because the thalamus is a primitive part of the brain, and both it and the insula are often considered centers of "automatic" sensation, beyond conscious control or thought. But Price's experiments—and those of other researchers, such as Robert Coghill of Wake Forest University[10]— suggest that effectively communicating that "things will feel better if we change" can produce a powerful range of assuaging reactions. In fact, expectations of relief can have a calming effect akin to a 6-milligram dose of morphine.

Financial advisers at moments of economic crisis have experienced this phenomenon firsthand. When they field calls from panicked clients, they routinely open the call by saying, "It is going to be OK. Let's not forget the big picture. Don't forget that we have prepared for uncertainties like this crisis. Let's stay focused on your values and what really matters." After reflecting on the fact that it is possible to navigate the storm, clients are more prepared to make the necessary counterintuitive moves, and advisers are more prepared to suggest them.

To take on a new narrative represents a leap of faith. It's not enough to see the value of the narrative or to desire change. You must consciously choose to instill new habits and new ways of looking at the world. You must believe they are going to make a difference, or you won't persist with them.

One of the most common types of business hubris is directly related to the problem of expectations versus desire. At a certain point in their history, many companies decide to become the largest companies in their category. This often seems to lead to missteps. Toyota and Volkswagen each declared their ambition to become the world's largest automakers, for example—just before they revealed operations scandals that were threatening to the whole company. The prevalent desire for continuous growth is understandable. It means the company will dominate its sector and control its destiny; many things are easier for companies that lead their industries.

But continuous growth for growth's sake, or for the sake of having everything be easier, is curiously divorced from a company's true self—from the things that set it apart from other companies. Perhaps that's why growth in itself (as in, "We really don't care how we do it, we just want to expand") doesn't work for most companies as a goal. It tends to be more effective to have a particular type of growth as a goal, along with a persuasive theory about why that type of growth is possible.

When you play the Wise Advocate role as an organizational leader, you too can use high expectations to advantage. To do this, you need to become the source of "free won't" in your company. Articulate the deceptive thoughts and the reasons why they have a hold on your company—and then suggest another way to interpret them. When you show that this new narrative is credible, and start to work on it, you foster high expectations. You can explain not just why a different direction should work, but why it will work: why the capabilities at your enterprise's disposal will get it to the finish line.

Reframing of this sort can change your life. One of the many people we've gotten to know who has experienced this is Sam Bonanno, a former senior executive at BHP Billiton, the leading global resources company. Before he retired, he had ascended to a job he had sought for years, a general manager position—in his case, leading and managing the distribution of mineral products to more than fifty customers in twenty-six countries. His group had more than four hundred people and almost $300MM in annual expenditures. He was also a member of the executive leadership team, reporting to the CEO of the global company.

But getting to this position had been fraught with tension. He had almost been passed over for the role, had seen it offered to others, and had been shunted aside into a job he considered a demotion.

"I felt betrayed and let down," he recalled. "I was devastated. I couldn't understand the logic, no one could explain it; there appeared to be no rationale." He considered leaving, and then he brought himself to a point where his Wise Advocate was evident. "I decided to work with what I was offered; I realized that this organization was still right for me in the longer term."[11]

But he also reframed his view of what was important to him. "I set a goal. I wanted to be the general manager at a mine, and if I played the right cards at the right time, the position would naturally come my way." He accepted the less senior role, and it turned out to be a stepping-stone to the mine manager position he wanted after all.

That wasn't the end of the story. After attaining the job he wanted, Sam became restless and increasingly aware of something "missing" in his life. His personal life was happy and balanced, but the professional landscape was no longer fulfilling a greater unidentified need within. He decided to take several months' leave and do the El Camino Santiago trek in Spain. He walked and walked for weeks. He carried a small notebook and pen in his shirt pocket and would pause, reflect, and write often. Gradually, he realized that deceptive brain messages had led him to spend decades of his life in pursuing goals that, in the last analysis, he didn't care that much about. Upon his return, he decided to leave the organization in which he had invested most of his working life and pursue a different direction as an independent consultant, a role that allowed him to devote a significant amount of time to the altruistic not-for-profit enterprises he had come to care about more.

You don't have to leave the organization to reframe its deceptive messages. However, by doing so, you will become more aware of your true interests—and those of the enterprise as well. And if you choose to stay, you may find your influence affecting the routines and habits of the larger system. Just as the brain has neuroplasticity—and can be rewired with an internal Wise Advocate coming to the fore—so does the organization. With the right new messages in place, people may no longer justify dysfunctional behavior by saying, "That's the way we do things around here," implying there is no way to change. Instead, they will recognize that they have already begun to reshape "the way we do things" themselves. Once they have reached that point, and others are working with them, then the organization begins to change its ways. By that time, you are in the step called refocusing.

CHAPTER SIX

Refocusing Your Attention

THE HEAD OF human resources at a manufacturing company, an individual who reported directly to the CEO, had a problem. One of the top salespeople in the company, a much-admired rising star with a great track record, had failed a drug and alcohol test. Officially, there was zero tolerance for drug and alcohol use at this company. Someone had leaked the information to a major customer, who had complained. The rising star had been dismissed, in accordance with company rules. The project manager (PM) had protested, but had been told the decision was final. Nonetheless, the PM had quietly hired back the rising star to jump-start sales for a new project. Now the head of HR had to deal with it.

Other individuals had failed drug or alcohol tests at this company in the past, but the cases were quietly covered up and never reported. This time, however, the customer's insistence on oversight had brought the problem to light; otherwise, it would never have reached the desk of the head of HR. And that was a symptom of a deeper problem that was holding back the whole enterprise.

At this company, as the HR director put it while telling the story, "the project manager is king."[1] For any project he or she leads, the PM controls all decisions, manages the flow of all information, and oversees rewards and recognition (including bonuses and promotion possibilities) for all staff. Project managers often developed a small core group of confidants, who would wink at what they saw as harmless transgressions, with no real consequence except breaking an unnecessary rule. The PMs also tended to operate at cross-purposes with each other, even undermining each other when they called on the same customers.

When the project manager hired back the rising-star salesperson, despite his dismal performance on the drug and alcohol test, the head of HR decided to step in. "A lot of people moved across to the PM's new project," he later told us. "Someone there decided that the person who had failed the drug and alcohol test should be given a second chance to work on this new project. The PM said he supported this and would give him another chance. I became aware of that plan through the safety team."

Now what? If the head of HR treated this matter in a heavy-handed way, with the stamp of authority, it would become a public clash. It might force the CEO to choose sides. No matter what the CEO decided, there would be a backlash. It would divide the company.

Remember the problem of virtue in a capitalist society, as articulated by Adam Smith, that we described in chapter 2? This is the organizational equivalent, a problem that any leader of a major enterprise must manage, often throughout his or her career: How do you give people the freedom to innovate and create while avoiding the temptations that lead to narcissism and exploitation? The head of HR at this company was facing a real-world example.

There were also, of course, deceptive organizational messages at play. The project managers were prone to misperceptions of risk and proficiency. *We don't have to play by the rules; drug and alcohol abuse won't affect us. And everyone knows that the project manager knows what's best.*

The basic problem here was not whether to rehire the rising-star salesperson or how to deal with an obstinate client. The head of HR wanted to establish the kind of company that could take this problem in stride. In the company he wanted to build, people would take problems like drug and alcohol abuse seriously, because they knew the effect that substance abuse could have on judgment and client relationships, but the policy wouldn't be enforced in a rigid, repressive way. The company would be the kind of mutually supportive, creative place where people chose to contribute to one another's success. Ideally, salespeople would choose not to drink or do drugs on the job because they wanted to stay sharp and mentally connected.

The head of HR recognized, as Adam Smith did, that it would not be possible to reorient the company merely through formal rules or laws. The project managers would simply continue to ignore the rules, bending them whenever possible.

Nor could the HR leader rely on the company's culture. "I recognized how the project culture works and how it is supported by other senior managers across the business," he recounted. "Instead of trying to fight every battle, I wanted to win the war." He had, in fact, discreetly checked in with a more senior leader about this, someone at the C-suite level. "This leader also held the view that the [rising star] should not be rehired," the head of HR later recalled. "But when he voiced his reasons to his peers, who were other senior managers in the organization, their reaction was to support the PM culture rather than the zero-tolerance policy of the business. This made me rethink what a wiser approach would be."

So the head of HR moved to refocusing. Rather than raising the problem publicly, he brought the Wise Advocate into play. He decided to initiate a series of conversations, one on one at first with key leaders, that would remind people, again and again, of the importance of a better reputation for the company.

First, he went directly to the project manager who had rehired the rising star. "I talked to him about the consequences of drugs and alcohol, and about sending the right message to others in the organization. It was not about stamping my authority, or taking power from the project that he is king of, but helping him glean a different perspective of what others may be perceiving of his project." In other words, he relabeled the misperceptions of risk as deceptive messages. What, for example, would happen if more clients learned of similar cases and the firm acquired a reputation for leniency in contradiction to its espoused values? He added that the company had an opportunity to lead in this area—to step out in front of their competitors with their visible skill and judgment.

Then he had similar conversations with other project managers and key individuals around the enterprise. Again, he didn't force compliance or flaunt his positional power as an argument. He asked about how difficult it would be to keep incidents like this from becoming public knowledge. He also talked about the company's competitive advantage, and the ways in which they were holding each other back. In talking about solutions, he kept reframing with the same phrase: *This could affect our company's reputation.* By using this message the same way each time, he drew everyone's attention, including his own, back to the downside of tolerance for drugs and alcohol.

In the end, the rising star was given another drug and alcohol test. When he didn't pass this one, he was once again dismissed. The story of the decision made its way around the company, and to many of its outside customers. The safety team, which had protested the rehiring, was reaffirmed. By this point, other project managers were buying in to the idea that it was time for the company to change. Other projects moved ahead safely. The culture of the company evolved further away from nepotism, and people became less tolerant of drugs and alcohol at work.

About a year later, another staffer in the same PM's organization failed a similar drug and alcohol test. This time, the staffer was a project supervisor's son. The supervisor pleaded for his son to be reinstated. The HR director simply went to the project manager and asked him what he wanted to do about it.

"The project manager decided it on his own," said the HR director. "He chose not to re-appoint this person. The values we are upholding are now being accepted as a rule for all, not a rule for some."

That is the power of refocusing. You are now playing the role of Wise Advocate for the enterprise at large, and thus reshaping habits that might otherwise seem incapable of change.

Refocusing Is Step Three

Let's say you've already made it through the first two steps. You've relabeled deceptive messages—of your brain and your organization. You've corrected the cognitive distortions, reframing messages into a new, more constructive narrative. Refocusing is the third step, in which you put this new, reframed approach into practice. You are, in effect, creating a change in your habits of mental activity—and in the related habitual behaviors of the organization around you—by returning your attention, again and again, to the mental activity and messages you have chosen to emphasize.

You are probably already aware of how difficult a change of habit can be. For example, there is the much-cited statistic voiced by Edward Miller, dean of the medical school at Johns Hopkins University. He was quoted as saying that among patients who have undergone coronary bypass or angioplasty surgery—an extremely painful and expensive procedure that only prolongs life for those who quit smoking, exercise, and lose weight—only one in ten, in his experience, changes to sustainably healthier behavior. The others see the value of doing this, but they don't follow through.[2] Similarly, people find it difficult to break free from obsessive-compulsive disorder or to overcome addiction. For many leaders, it is so strikingly difficult to shift and transform the direction of their organizations, that comparisons to overcoming OCD and addiction seem apt.

And no wonder—for in a sense, the same dynamic is at play. The missing link is refocusing—an intentional practice of conscious repetition with awareness, deliberately reorienting your attention in chosen directions. This practice sets up new patterns of mental activity that rewire the brain in some powerful new ways. First, you are consciously constructing a new habitual pattern of consulting your Wise Advocate: *What is the best decision to make in this case—the decision that a dispassionate, well-regarded observer would respect?* Second, you are shifting mental activity from the Low Ground to the High Ground (and, ultimately, to the Higher Ground). Third, you are creating a new pattern of mental activity that rewires the brain, building direct connections between the Executive Center and the Habit Center.

For many people, this represents a major shift. Typically, Low Ground activities become habitual because the propensity to repeat them is so strong. As we saw in chapter 2, this is generally because the Low Ground brain circuit connects the Habit Center to two other powerful centers of brain activity: the Reactive Self-Referencing Center, associated with concepts of identity related to subjective valuation (*Is it valuable and relevant to me?*), and the Warning Center, associated with the sense that something is wrong or action needs to be taken. The Habit Center is the source of many feelings we associate with habits, including pleasure and familiarity—"doing what feels right."

Because your mind has most likely traveled the Low Ground a great deal in the past, with only sporadic consultation of your Wise Advocate, your habitual response to a challenge will often be to do something expedient; often that means paying attention to the deceptive brain messages that circulate on the Low Ground.

The High Ground also connects to the Self-Referencing Center (but a critically different part of it, the deliberative part) and the Warning Center. But it does not primarily connect to the Habit Center in the same way that the Low Ground does. Instead, it is more closely related to the Executive Center—the lateral prefrontal cortex, associated with adaptive, forward-focused mental activity and the cognitive processes of executive function that are so important to the Wise Advocate. In refocusing, as you bring your attention again and again to the High Ground, you leave behind the old Low Ground mental habits. You choose new habitual patterns of mental activity, those associated with executive function, with more long-term benefit and more value to your true self than the habits associated with the Low Ground. In this way, you retrain your Habit Center to serve the High Ground rather than the Low Ground.

To accomplish this, you must return repeatedly to the High Ground, reinforcing that pattern of mental activity, strengthening the brain circuits related to it, and bringing the Wise Advocate into play each time. In other words, the Habit Center starts to cue you to consciously bring attention density to bear. As you may recall from chapter 2, attention density is the practice of focusing your attention in a consistent way, over minutes, hours, days, weeks, months, and years. Attention density can be thought of as developing high-intensity and (potentially) high-resolution observational power—the ability to observe more in the same amount of time. The more you focus your attention on a specific idea or mental experience, the higher the resolution you create around it.

Sometimes people assume that attention density means concentration—a concentrated focus of attention. But attention density doesn't have to be laserlike in its sharp focus; it has to be consistent and repetitive. You can include a range of narratives and thoughts, but you have to return to them, and they have to complement each other and move you in the direction you want to move. When you focus your attention on them repeatedly and (better yet) self-consciously, with awareness that you are doing this, it accelerates the rewiring of your brain. (As we'll see later in this chapter, this acceleration of Hebb's law is known as the quantum Zeno effect.)

Similarly, at the larger scale of the organization, refocusing shifts the tone of conversations and thinking. Calling on the Wise Advocate is no longer something unusual; it becomes, over time, simply "the way we do things around here."

The eminent executive coach Marshall Goldsmith credits this kind of refocusing (though he doesn't call it that) as the primary source of his success. During his lectures about leadership, Goldsmith asks people to consciously change some aspect of their own behavior at work, such as the way they conduct meetings. He asks them to explicitly discuss this improvement priority with their coworkers and subordinates. Then he asks them to follow up by checking in with those individuals: "How am I doing?" There is also a subsequent step: "I want you to come back and email me, Marshall Goldsmith, about how you're doing. I want to know."[3]

The results are remarkably consistent, not because of the change itself, but because of Goldsmith's explicit request that they close the loop with their fellow employees and then with him. The attention that this requires is discomfiting at first, but it moves the activity of initiating refocusing into the executives' Habit Center. Leaders who continue to refocus in this way, following through with subordinates, asking them every few days how well they are doing, will find that the process triggers everyone's Wise Advocate—their own, their subordinates', and even that of others around them. Something similar takes place with the right kinds of after-action reviews and postmortems. While they are discomfiting at times, they bring explicit attention and the Wise Advocate to areas that could otherwise become deceptive organizational messages.[4]

Goldsmith himself practices the same type of discipline—but even more stringently. Every night, he forces himself to do one of the most disciplined things imaginable. He has a friend act as a Wise Advocate, calling him and asking him the same twenty-two questions. (Remarkably, they are still friends.) These questions all start with the phrase "Did I do my best [today] to"; the endings may be strategic (". . . set clear goals?"), professional (". . . preserve all client relationships?"), philosophical (". . . be grateful for what I have?"), physical (". . . exercise?"), or personal (". . . say or do something nice for Lyda [his spouse]?"). Many of them are directly related to increasing his own leadership skill: "Did I do my best to learn something new? Did I do my best to avoid destructive comments about others?"

He has done this for years, revising the questions as his priorities evolve. It's noteworthy that he doesn't just answer; he rates his efforts that day for each question on a scale of one to ten. If he's honest with himself, and there is every reason to believe he is, then this is, in effect, a dramatic form of refocusing; he is paying intensive attention to the gap between his aspiration for the day and his actual conduct, with a view to bringing himself closer to the goals of his true self. From our perspective, he is systematically connecting with his Wise Advocate.

The questions are one example of what Goldsmith calls "triggers," or deliberately designed environmental cues that move you continuously, relentlessly, in the direction of productive, beneficial change. "My mission is to help people become the person that they want to be," writes Goldsmith, "not tell them who that person is." He adds that this is the only way to get results, or to generate serious change. In other words, he believes that if you don't put yourself through a regimen of self-scrutiny—if you don't spend some time focused on how well you followed through on your promises, including the promises you made to yourself—then you will not realize your potential as a leader. (This is, in our view, very similar to inner dialogue with the Wise Advocate.)

The momentum generated by this kind of focus is hard to believe until you see it at work. A small group of people within an organization, continually focusing the group's attention on a few messages that draw on people's Wise Advocates, will typically have far stronger ongoing influence than a much larger group, higher in the hierarchy, trying to change the entire culture in a wholesale way.

Focusing your attention in this new way, over and over again, is just as difficult as reframing was. Refocusing is like a training regimen for your own mental habits; it invokes attention density, mindfulness, and mentalizing. At high-stress moments, it involves presence of mind: deploying the veto power of "free won't," stopping your momentum long enough to choose the reframed approach. And in your low-stress moments, you have to practice a reframe-oriented line of conversation, so that you already have the habit developed, at least in part, ready to execute when you are under more stress, and so that people around you recognize your consistency. You are consistent in your message, and you return to it when you are distracted.

At first it may feel as though nothing is changing, but the power is in the focus: the repetition of reframed messages will, sooner or later, have an effect. Gradually, your behavior will shift. Instead of reacting to each impulse with immediate action, you will consider whether it is good for you in the long run, and you will think about feasible alternatives. Thus you will gradually change yourself into the wise leader you need and want to become. Refocusing changes the brain by leading you onto the High Ground.

The Quantum Zeno Effect

Many people don't think that they can change themselves—they need outside help. Outside help can certainly make a difference; we've seen the benefit of an experienced coach or guide. But the greatest accelerator of attention density is a phenomenon we call meta-attention, in which you pay more conscious attention to the way you are shifting your attention.

For example, imagine thinking *I am less distracted now than I was just a minute ago.* Now you have a mental image of yourself, about to pay attention to a message that you articulated in reframing. If you then pay attention to that message, right after you imagine yourself doing so, you invoke the quantum Zeno effect (QZE), which makes refocusing all the more powerful.

The quantum Zeno effect is a principle of physics that explains the value and power of focused attention. Neurons communicate with each other through a type of electrochemical signaling that is driven by the movement of ions such as sodium, potassium, and calcium. These ions travel through channels within the brain that are, at their narrowest point, only a little more than a single ion wide. This means that the brain is a quantum environment and is therefore subject to all the surprising laws of quantum mechanics. The QZE was described in 1977 by the Nobel prize–winning physicist George Sudarshan at the University of Texas at Austin and has been experimentally verified many times since.

The QZE is related to the "observer" effect of quantum physics: the nature of an observation affects the way the system behaves, even if the observer doesn't directly intervene in any way other than by observing. In physics, the observed behaviors of subatomic particles, and the behaviors of larger systems built out of them, are expressed in terms of probability waves. When any system is observed in a rapid, repetitive fashion, it affects the probabilities of change, and the rate at which that system changes is reduced. One classic experiment involved observing beryllium atoms that could decay from a high-energy to a low-energy state. As the number of measurements per unit time increased, the probability of the energy transition fell off: The beryllium atom stayed longer in its excited state because the scientists, in effect, repeatedly asked, "Have you decayed yet?"[5]

In the mid-2000s, physicist Henry Stapp at UC Berkeley applied the concept of the quantum Zeno effect to an understanding of how attention can change the brain. Stapp and Jeffrey Schwartz argued that the act of focusing close attention on your mental experience—whether a thought, an insight, a picture in your mind's eye, a pain, or a feeling of relief from that pain—stabilizes the brain state arising in association with that experience. Over time, paying enough attention to any specific mental experience strengthens the firing patterns of the relevant brain cells and brings Hebb's law (cells that wire together fire together) into play. These brain connections can eventually become not just transient links but stable, physical circuits, resulting in actual changes in the brain's physical structure.[6]

There are also signs that through attention density, the veto power of "free won't" can be strengthened, by strengthening the requisite brain circuits (those associated with High and Higher Ground). Through sustained observation (of which meditation is a cardinal example), the executive brain function is enhanced, resulting in enhanced capacity to inhibit behavioral responses. With enhanced attention density (more observations per unit of time), you observe reality more intensely; more attentional experiences in, say, a one-second period means that your awareness of the world (and of your thoughts) and your mindfulness potentially have a higher resolution. The increased capacity of your attention is like a shift from a low-resolution photo to a high-res photo. This can increase your ability to shift brain circuits, because it stabilizes the connections within the circuits associated with the mental activity you have focused on. It causes those High and Higher Ground brain circuits to wire together, while other brain circuits, associated with mental activity that is not focused on, become less connected.

This ability is particularly enhanced by a strengthening of the ability of your Executive Center to inhibit responses to impulses. There is very good evidence that practicing focused attention enhances the function of the Executive Center,[7] and that both inhibiting behavioral responses and cognitive flexibility are key executive brain functions.[8] (As we saw in chapter 3, this is correlated with the High Ground.) The more practiced you become at reframing your thoughts and choosing the direction your attention will take, the easier this becomes.

One leader we know who has experienced the quantum Zeno self-directed neuroplasticity effect directly is Sandra, the head of human resources for the local office of a large transportation company. She reports directly to the head of "people and culture," who reports to the CEO. She manages recruitment, remuneration and benefits administration, office administration, and employment conditions.

Sandra's relationship with her boss was often problematic. Deceptive brain messages tended to emerge for her when her boss was busy. "There are times he'll spend a lot of time with me, and other times when I'll have minimal contact. When this happens, I would dramatize and react in my head by questioning: *Am I a valued member of the team? Does he even want me here? Maybe I don't matter! What are my future prospects like?* My reality would be consumed with feelings of fear, stress and anxiousness."[9]

She went through the exercises in relabeling and reframing, but she didn't really gain the confidence to act differently. Her reframed message went something like this: *It's a good thing my boss is not spending time with me right now because I can just get on with my job; no one's looking over my shoulder, I have autonomy and freedom to get my work done!*

At first, this was a rote exercise for her; each time she felt as though she was going through the motions. But at a certain point, she started paying attention to the way she was paying attention to this. "I'm not really taking this seriously. Part of me still thinks my boss is ignoring me because he doesn't value me."

Now she was aware that she had two different types of mental narratives. She began to reflect on why she felt both were valid, even though they contradicted each other. What were the reliable signals from her boss (and the rest of the organization), and how could she more accurately interpret those signals?

After a month or two, she noticed that she was not nearly as troubled as she had been. It had become a habit to notice which of the two narratives she was paying attention to, and why. She began to feel, at a more visceral level, that there was no direct cause for concern. These messages were thoughts, not facts.

"The working relationship is sometimes like the sun shining, and sometimes it's like the lines to authority are hidden by clouds," she said recently. "I know I'm still getting the support I need even during the cloudy periods. I have refocused things so that whenever I feel I'm being cut off, I feel like I've been given more freedom and autonomy. Within a few weeks, our contact will recur. In the meantime, I can just do my job, and cultivate my relationships with everyone I work with, not just my boss. It has made me a stronger leader overall."

Refocusing at Cargill

The quantum Zeno self-directed neuroplasticity effect can also occur at an organizational level. For example, at the Cargill agricultural and food products company in the early 2010s, a deliberate use of meta-attention helped spur a change initiative forward. The very name of the program, "Constructing a Creative Narrative," suggested that it involved more self-awareness by participants than the typical corporate change program does. To buttress that further, the leaders of the company openly declared that they had designed this program to change behavior up and down the hierarchy, and that they were using principles of neuroscience to help people shift the way they operated.

Cargill had already undergone another major shift, starting in 1999, toward becoming a less bureaucratic, more agile, solutions-based organization. The company's executives had issued a statement of values, defining the "heart of leadership" for their company as integrity, conviction, and courage. They had also set out to create a "culture of freedom," empowering and encouraging employees at every level to act with decisiveness and authority on behalf of customers. This was a familiar move for any company that wanted to compete.

But as with many such programs, the "heart of leadership" initiative never fully took hold. Even after more than a decade, the company was still not as customer-focused as it needed to be. One customer, a large packaged-foods manufacturer, told a Cargill executive, "You send fifteen different people to our offices each week from different businesses, and they all ask us some of the same questions, but they never try to understand exactly what we do with all of your ingredients. If you brought all those people together, you could potentially offer much more to us."

The situation clearly called for new behaviors. Better collaboration among Cargill employees, for example, would not just solve the problem of redundant sales calls; it could also lead to new logistics, risk-management, and quality-assurance practices. But that type of collaboration, especially across Cargill's seventy-plus businesses operating in sixty-six countries, would be a stretch—particularly since in Cargill's culture, it would require broad, widespread commitment.

In 2006, the company's leaders met to figure out how to move forward. They articulated some major behavioral, structural, and cultural changes that were needed—in effect, a major shift in "the way we do things around here." The resulting initiative began with promoting a change in leadership style. As David Larson, then the executive vice president (since retired) put it, "Our good leaders are those who focus on others, give undivided attention, and build trust. Leaders can either give energy to people or drain energy from people." Many leaders within the company instinctively knew how to translate this into their own day-to-day behavior. For others, including some who had been at the company for fifteen years or more, this concept required a major shift.

The economic crisis, which began the following year, gave the Cargill leadership a chance to practice that shift. The Cargill leadership developed a couple of new messages that became symbols for how they would manage through the downturn. They set up a cost-management effort, designed to reduce expenses, and called it "hunkering down wisely." The phrase, repeated again and again, symbolized confidence. *These cuts will make life better for the company (and its employees) in the long run.* To deliver on this reframed promise, the company could not cut costs across the board. Instead, the leaders moved strategically, looking at each cost as a potential investment and cutting thoughtfully, where it would yield the most benefit. The idea of hunkering down wisely helped calm anxiety about Cargill's ability to weather the crisis, and it empowered people to come up with creative ways to save money for the company without short-changing its ability to operate. It led to a far greater sense of ownership and effectiveness than would have been produced by across-the-board budget cuts or other top-down directives.

One sign of the value of this message, and the way it helped the company refocus, was the way the name of the program, "hunkering down wisely," was kept alive. It even jumped into the local night-life scene when an assistant treasurer at Cargill named Dave Braden formed a band, with musicians whose ages ranged from twenty-five to fifty-eight. They named it the Hunker'd Down Blues Band. Within a year or two Cargill was profitable again; meanwhile, the band continued to perform and record.[10]

Another message emerged in executive training, where Cargill's leaders were taught to deal with difficult situations by classifying them in one of three ways: problems, predicaments (impasses), and polarities (situations with conflicting goals). "If it's a problem, we work on solving it," explains a Cargill executive. "If it's a polarity, it's not an 'either-or' situation but an 'and' issue that requires management. And if it's a predicament, you have nothing to solve or manage; you can only accept and endure."[11]

Leading a Refocusing Initiative

To lead a refocusing initiative, you must be able to draw on the Wise Advocate. In our experience, four key factors come into play. The first is focus (attention density) and how it influences brain function, which we've already discussed in depth. The other three are goals, commitment, and action.

Goals

In the new, reframed messages, as we saw in chapter 5, you focus on the future: what you are trying to create. The Wise Advocate in your mind will probably frame this differently than you would if you were on the Low Ground. It will frame your goals and aspirations in a way that moves the entire enterprise forward on the High Ground. Focus on what you want, rather than on what you don't want.

Thus, in the story at the opening of the chapter, the head of human resources focused attention not just on finding a solution to the problem of one employee, but on the ongoing reputation of the entire enterprise. And even then, the reputation of the enterprise depended on the entire company taking its customers seriously and caring enough to stop the common practice of drinking and drugs at work. Not just making promises but keeping promises—this is how trust is maintained.

Similarly broad aspirations have been a component of every story we've told. In the Alcoa story in chapter 1, Bill O'Rourke didn't just set a goal of safety; he set a goal of creating a thriving enterprise in an otherwise authoritarian environment. Aetna CEO Mark Bertolini didn't just set a goal of giving his employees a higher wage; he set a goal of becoming the kind of company that can play a leading role in the new health-care environment in the United States. Queensland Investment Corporation's leaders didn't just want to bypass the intermediaries who managed their portfolio; they wanted to find a new way of investing with integrity. Cargill's leaders, similarly, wanted to build more effective customer relationships, in the service of becoming a better company.

Some people believe that work is purely transactional—that people simply do a job in exchange for pay and benefits. But refocusing really gains force when people's personal aspirations align with their job in a way that makes them think about the value of both, time and time again. In his book *The Fifth Discipline*, Peter Senge shows how private aspiration can be linked to company performance. He argues for a practice of "personal mastery" tailored to whatever level of aspiration the individual wanted to articulate.

"Vision is multifaceted," Senge writes. "There are material facets of our [aspirations], such as where we want to live and how much money we want to have in the bank. There are personal facets, such as health, freedom, and being true to ourselves. There are service facets, such as helping others or contributing to the state of knowledge in a field. All are part of what we truly want."[12] For each individual, he goes on to say, the balance of goals is different; and part of the discipline of personal mastery is the continual development of the skill of knowing what you want and why you want it. As you become more aware of the reasons why you want to reach your short-term goals (those associated with the Low Ground), you reorient them toward longer-term aspirations, like those associated with the High Ground and the Higher Ground.

Under every short-term goal, in Senge's view, there's a longer-term aspiration. For example, even if you want more money right now, why do you want it? To feel secure? To have more status? To purchase certain goods? To have a different type of experience? To provide for your family? To solve some problems? If you were able to achieve what you want, what would that get you? And if you achieved that, what would be the result? If you keep asking that question, it will ultimately lead you to questions about your ultimate aspirations and your true self—questions related to High Ground mental activity.

One piece of support for this comes from Jamil Zaki's research on empathy. He found that thinking about what you want or what other people want is closely associated with some of the brain circuits we call the Low Ground; having a broader, nobler altruistic aspiration is linked to what we call the High Ground.[13] This makes sense when you consider the impartial-spectator aspect of the Wise Advocate: seeing yourself as others might see you. From that perspective, the expedient outcomes—pleasing yourself, pleasing others—will no longer seem so rich or appealing.

Commitment

As we noted in chapter 3, the major difference between Adam Smith's impartial spectator and the Wise Advocate is your connection with it. The Wise Advocate cares for and loves you—and expresses that care and love in your mind. We alluded to this in chapter 5, when we talked about the expectations you have in reframing your messages. We said that expectations matter: it is important to expect success. One reason for this, we think, is the link between expectations and commitment. If you don't believe you can change, it is all the more difficult to care. Paradoxically, though detachment is important—the Wise Advocate must be able to observe dispassionately, like a truly impartial spectator—it comes alive when combined with commitment. You look on with an outsider's perspective, but you care about the results, even more than you did before.

The act of refocusing requires massive commitment. You cannot do it unless you truly care about yourself and your organization. Day after day, conversation after conversation, you return to the same ideas, focusing attention on them again and again. A focus without that kind of commitment will be short-lived and transient; it will float away. Indeed, this is much the same type of commitment that someone makes when diagnosed with a serious illness. The treatment that is prescribed to cure it may or may not work; but the healers know it will not work unless the patient is fully committed to recovery.

Commitment depends on the willingness to change and take responsibility. If you are a steady smoker, and you learn that a close friend has passed away from lung cancer, you may feel enough agony and grief to stop smoking for a week or two. But that, in itself, will probably not provide the commitment you need to refocus for the long term. You need to viscerally recognize the harm that this habit does to yourself and those you care about. The same is true for making any significant organizational change.

Paradoxically, one of the barriers to commitment may be your own sophistication as a leader. You may already be keenly aware of the deceptive brain messages in your mind. You may, in fact, like them. Comfort zones are comfortable. You may know that with "free won't," and with attentiveness to your own attention, you could move to the High Ground and leave those messages behind—but you can't quite bring yourself to do it.

This is called impulse loyalty. You become reluctant to abandon the familiar cravings and thoughts that comfort you. You may like the familiar feeling of being on the Low Ground.

"I can get away with it," said one such individual, who was taking chances in charging personal expenses back to the company. "I'm going to gamble that I'm going to be one of the lucky ones and if I can't, then I will just take my lumps and chalk it up to bad luck."

Another said, "If I had known I had to refocus my preferences in order to take this new job, I would never have agreed to it."

And still another said, "I've come to resent the conflict between work life and home life. My solution is not to deal with it. I'm just spending more time at work."[14]

No one can argue with these perspectives. We could tell you they represent a form of narcissism. We could compare them to misperceptions of risk and proficiency, two categories of deceptive brain and organizational messages that we described in chapter 4. These messages tell you that you've been sharp and competent enough to get away with this kind of thinking so far. You are special enough that you can remain loyal to your impulses without suffering many consequences. This problem is even more pronounced if you are already in a senior leadership position, with people constantly reinforcing the idea that you are special and telling you what you want to hear.

Impulse loyalty is a habit. It can last for a long time—for as long as you feel you can keep it up without too many negative consequences.

Sooner or later, however, as with someone treading water, you may reach the limits of your endurance. You will not be able to make the moves you want to make. You will not be able to realize your aspirations.

Given the persistence of impulse loyalty, the ways in which many organizations reinforce it, and the difficulty of overcoming it, how can you as a leader reading this book master this problem? There's no glib answer. There is no universal scientific approach to resolving it, for no two brains are exactly the same, and every solution must be situation-specific.

And yet, as the concept of Free Won't suggests, there is reason to expect success. One answer is to work at a micro scale, looking at immediate impulses and associating them with disgust, one by one. Alan Carr, writer of a series of successful books on how to quit smoking, relied on this method to help people drop the cigarette-smoking habit. (Carr himself, who had given up a hundred-cigarette-per-day habit at age forty-eight, died of lung cancer at age seventy-two.)[15] One of the things that makes the habit appealing is the idea that smoking provides pleasure—and a relief from uneasiness or stress. When stress strikes, a deceptive brain message appears: *I'd love a cigarette right now.* This leads to a longing for that feeling of taking the first puff.

To reframe this requires some mindfulness: you choose to become more keenly aware of how pleasant it feels to not be smoking, and how truly miserable it would feel to draw the poison of tar and nicotine in your lungs. Having reframed this feeling, you can then refocus: you remind yourself, again and again, of that second feeling, mentally reliving the disgust and unpleasantness that is also involved in the act of smoking. When a moment of stress or uneasiness comes your way and you feel like having a cigarette, you can call upon the reframed feeling of disgust, because you've rehearsed it and it is available to you. And then you learn that if you abstain, the feeling of uneasiness goes away on its own. *This craving is not permanent. All cravings pass.*

Every moment of refocusing involves a choice about where you will place your attention. Often, you will need to choose a different pleasure than your impulses are oriented toward, but an equally powerful one—one that is good for you and to you. If you are trying to counter a craving that is bad for you, all of the standard suggestions apply here: taking a walk, taking a nap, meditating, cleaning up the room, writing about the craving in your journal. When you refocus, you are aware that you are doing this explicitly to combat your craving. You know exactly where the deceptive brain message is taking you, and you expect that you will be able to counter it.

Yet there will be surprises. Any refocused message will operate in subtle ways that you can't exactly foresee and that will continue to ripple out through your life. Illian Ivanov of Mt. Sinai Hospital has found, in working with children, that these almost unnoticeable interventions, if repeated, are very powerful in changing habits. A single intervention—a conversation, a medication, or a single time avoiding an addiction—has limited power. But over time, he says, the interventions add up. You may have a conversation with someone that seems to have little or no effect—but find out, a year later, that everything has changed for that person. Commitment, in this case, didn't come from any one emotional moment of truth. It built up bit by bit over time.[16]

The practice of mindfulness may make it easier to make the commitment—or at least to realize what kind of commitment you are willing to make. If you can find the Wise Advocate in your mind, you may realize that you are more interested in making that commitment than you thought you were at first—and that you have the resources to do it.

Action

Leaders focus others as well as themselves. In this way, they translate the new messages, and their own effort at refocusing, into events that affect the world around them.

Consistent action leads to further repetition. When you treat customers in a new way, or conduct meetings in a more measured way, or commute by bicycle instead of by car, you continue to refocus. Each little change is like a "small win," and each one is contributing to your shift toward the High Ground. A new habit is formed through intentional, focused repetition and positive feedback/reinforcement.

There is still much to learn about how to translate attention density into action, especially when it is self-directed. For example, some chronically late people have learned to carry two timepieces—one fast and the other accurate. Why does this help them? Could it be because the gap between the two timepieces forces them to focus attention on their own schedule, and this, over time, makes it easier for messages favoring promptness to emerge?

In 2008, General Motors used the link between refocusing and action to bounce back from bankruptcy and to pave the way for repaying the U.S. government bailout it had received. CEO Ed Whitacre and Mark Reuss, the president of General Motors North America, decided to reduce the rebates that dealers received to give them the incentive to sell cars. Rebates represent an almost irresistible impulse for many automakers, who must keep their relationships with auto dealers strong. If sales drop, automakers can provide two to five thousand dollars for dealers to use to reduce prices.

Unfortunately, rebates are not just drains on profitability; they lead everyone, from the customer to the automaker, to attribute less value to the car. They are like a tangible manifestation of the destructive deceptive brain message *Our product isn't worth what we charge for it*. They are also capriciously bestowed on customers who complain, thereby reducing brand loyalty, which automakers depend on in the long run. A company that offers rebates to customers who complain about the price is signaling that they don't really care about customers unless they complain.

So Whitacre and Reuss stopped the practice. This required relentless effort. The impulse toward rebates had been embedded in the GM system over the years. Whenever it came up, at any level of the company, there was now a clear mandate to stop the impulse—to exercise "free won't." To make this work, Reuss had to demystify pricing, setting up clear guidelines for prices, where before they hadn't existed, and dropping GM's average rebate by $1,200 per car. As Bloomberg writer David Welch recounted in a 2010 article, "GM's market share tumbled to 17.6 percent in March from 19.4 percent during January and February. But Reuss kept prices up and beat his sales goals."[17]

Refocusing can also involve small changes in personal actions. An executive who said he was too busy has now set aside one hour per day for self-directed activities aimed at his own health: a small amount of exercise, cooking, or time spent with a child or friend. He has to consciously give himself a mandate to focus his attention during this hour on himself. His deceptive thinking had focused on his commitment: *If I take time off, I'll remind people that I'm not dedicated enough to handle everything they throw at me*. Now, in refocusing, he reminds himself that the real issue is not his commitment, or his competency, but his capacity: *I need to allow time for self-care so that I can give people what only I can provide them*.

Perhaps you are thinking, "This all sounds too easy. Is the answer to all the challenges of change just to keep returning people's attention to the same few points of a reframed narrative?" Apparently, that's what works consistently. And some of the most successful management change practices have this type of principle ingrained in them. "Open-book management," for example, has been credited with remarkable gains at companies like Springfield Remanufacturing Corp., because it repeatedly focuses employees' attention on the company's financial data. SRC is a consortium of companies, all based in Springfield, Missouri, all employee-owned, mostly devoted to rebuilding or reselling industrial components, and they make their financial-performance numbers freely available to employees and staff. A profile of company founder Jack Stack describes the way the system changes the thinking of every employee, through continual refocusing:

> Everyone, including visitors, knows exactly what the score is, because the company's financial information is posted on mural-sized charts in the employee cafeteria. Staff members "huddle"—a word deliberately chosen to evoke game-playing strategy—at least weekly to talk about the financials and the strategies they suggest. . . . Megavolt General Manager Dianna Devore, who started at SRC 16 years ago as a clerk/typist, and Jack Stack, who casually drops in to visit Megavolt, have both been known to pull visitors over to the charts and explain some significant shift in overhead costs or cash flow.
>
> Since the charts get updated daily, people know very early whether they will hit their monthly targets, and why. "That's the key," Ms. Devore says. "If a plan projected $332 in overtime, and we see we're paying out $351 halfway through, we know that we must react. Something is happening. If we waited for an accounting person to tell us, it would be too late to do anything about it."
>
> [Open-book management] provides the missing direct experience of the business specifics, in a way that people have reason to trust, because they know their bosses are looking at the same numbers. . . . One of [its] most important [assets is] the years of experience with financial literacy, disclosure, and close-to-the-edge financial operations. If the . . . businesses start to fail, it won't take SRC workers by surprise. They'll know well in advance, because they will have been tracking that failure, huddle by huddle, on the wall charts.[18]

Refocusing is triggered by the repetition: Friday lunch after Friday lunch, talking through the numbers, until the understanding implicit in them is embedded deeply in the way people think and act. One result is SRC's ability to routinely invest in new enterprises launched by its own employees, including many without a college degree. More than half of these businesses have become viable long-term enterprises in their own right, a remarkably high success rate. That's because the open-book system has given all SRC employees the pattern of thinking they need to run a company themselves.

Toyota's production system, similarly, involves people at every level of the company in developing a fine-grained awareness of their processes and how to improve them. In both of these approaches, in workplace sessions that occur weekly or even daily, people systematically talk about the means for making things better, training their brains to make new connections, returning again and again to the same questions and methods of improvement.

Even the act of reading about refocusing is itself a kind of refocusing. By thinking about whether you are going to focus attention on deceptive brain messages, you are reinforcing the slow-moving response within yourself. You may not think you're reacting, but by reading books like this and having conversations like this, you may start to see the world differently. It may take some time before you realize, like many people before you, that in priming yourself for change, you have already begun to shift your habits. When that shift is underway, you're ready to take your leadership to another level—and thus to begin the step of revaluing.

CHAPTER SEVEN

Revaluing Your Leadership

SOME EXECUTIVES WHO rise rapidly to senior positions experience panic and anxiety attacks. As we said in chapter 1, the challenges they face are complex, many-faceted, "adaptive" or "wicked" problems, difficult to manage in any rote or formulaic fashion. To address these issues, the leaders will have to take the organization—and themselves—up to a new level of awareness and competence. They are not sure they can do it. They feel they cannot talk about this openly because it would make them look weak or vulnerable, and that would make it even more difficult to succeed. This makes them more likely to ruminate over the deceptive messages in their minds: *Now that I've reached the top, I can't trust anyone to help me.* Or, *I have to act like I have all the answers now.*

Given the "leadership contagion" effect that we mentioned in chapter 2—the prevalent habit many people have of adopting the attitudes and values of visible leaders—these cognitive distortions are likely to carry over to the rest of the enterprise. If the CEO and other prominent leaders believe, for example, that the ordinary rules don't apply to them, or that they can't take a chance on any risk, everyone else in the enterprise will act accordingly. That's how organizations get addicted to self-destructive behavior.

The stakes are high, and not just because your own job may hang in the balance. These days, everything that happens of importance in civilization happens through organizations. Your organization is no exception. The step of revaluing is where you enter the fray whole-heartedly, putting into practice what you have learned in the first three steps, but now at a larger scale, outside yourself. You do not automatically accept the habits and routines of your organization; nor do you automatically reject them. You set out to rethink their value, not all at once but bit by bit. You do not act alone, but in collaboration with like-minded colleagues who recognize the same issues you do, and who will collaboratively work with you to help bring the organization where it needs to go.

You are aware of one thing that others may not see: decisions made while influenced by the Low Ground—no matter how beneficial in the short run—tend to undermine your enterprise's ability to reach its long-term potential. The Low Ground can help your enterprise make deals and satisfy immediate wants and needs, but it can't help an organization transcend its limits successfully. For that, it needs strategic leaders: people who can play the role of Wise Advocate in the organization, echoing the voice they have cultivated in their own mind, but now at a larger scale.

In this step, you practice that role and become proficient at it. You will probably spend much of your time trying to reframe the deceptive messages that are rampant in your company. Your reframed narrative will often take this form: *This message is not the full reality; it is simply what we have been telling ourselves. There is another way of looking at it.* You will feel called upon to articulate other messages that can lead the enterprise in a better, more clear-sighted direction. Indeed, this is what the ability to "speak truth to power," in our opinion, really means.[1]

We call this step revaluing to emphasize that it is not a neutral position: you are taking on genuine advocacy, on behalf of the organization's "true self." You are also exercising judgment. To value something means to assess its worth—to come to an understanding of the contribution it can make and the respect others will hold for it. To revalue means to reassess, to look with broader perspective at your own leadership—in this case, in light of the understanding you have gained through relabeling, reframing, and refocusing on new messages—and to use that reassessment to create a better context for your work and for the organization around you. To revalue is to translate the inner voice of the Wise Advocate in your mind into the kind of judgment that can help the organization make the right decisions. This kind of strategic leadership is characteristic of those who can step back from the transactions of the moment and occupy the High Ground of their mind.

Here are two examples of people who have done this, from the many we have known over the years. Both are people in relatively senior positions who could have limited themselves to meeting their targets and making expedient choices. Instead, they chose to have more impact and help an organization transcend its limits. They didn't do this on their own; they worked closely with their chief executives and other senior leaders. Their influence came only in part from their positions. It came more from the way they conducted themselves, as a catalyst for helping people throughout the organization see themselves more clearly. In the process, they made it possible for their organizations to cross a necessary threshold.

QSuper: Commerciality with a Heart

Government-owned investment funds are not generally thought of as innovative enterprises. Yet their role in guaranteeing pensions and prosperity requires them to continually raise their returns, and if they don't, they can lose their protected status. In 2016, that happened to QSuper, one of Australia's largest annuity funds and the chief finance organization for pensions in the province of Queensland. The provincial government no longer required government workers to use QSuper, and it thus found itself competing with the private sector.

Like QIC, whose story we told in chapter 6, QSuper's culture had been shaped by its status as a state-owned financial services firm. It was (and is) a 1,200-employee firm which oversees about A\$75 billion in assets (about US\$55 billion) for about 560,000 customers. And like many state-owned firms, it had leaders who were eager for change. The incoming CEO Michael Pennisi, who took over the top job the previous October, had previously been the chief strategy officer and had seen the change coming for a while. He also saw a solution: to enter the private-sector financial world wholeheartedly, competing with private enterprises in the pension space.

QSuper was already known for its inventive ways. Under Pennisi's predecessor, Rosemary Vilgan, it had embraced digital technology and established itself as a leading financial player in Australia. But the firm was still bound by its past sense of purpose as an administrator of funds for pensioners, overseen by the state. Now it was, in effect, defining a new way of providing value to its pensioners.

"We had to up the ante on commerciality," Pennisi says, "if only to make better profits for our members. But we also had to keep the best of our existing culture. When a company undergoes change, it is important to retain the core values that have built trust with customers, employees, and the larger community. QSuper's strong financial position and member relationships are the essence of our business."[2]

Pennisi reframed the company's purpose, using the phrase "commerciality with a heart." He articulated a new type of management process, which he called "decentralized decision making," essentially empowering employees to conduct more of QSuper's business in an entrepreneurial way, without relying on top-down direction. This was necessary to build the levels of customer service that QSuper needed. Pennisi put in place an organizational structure to support this type of entrepreneurial spirit and devoted himself to refocusing attention, again and again, on the new ways of operating. "We invested a lot in our people, in our leadership capabilities, and in being very clear about what we stand for, what we expect of each other, and what we will accept."

There was a lot of conversation about this, first with top executives and then with employees, placing attention squarely on the decisions the company was making—including tough decisions about layoffs. "We're going to talk about these decisions openly and why we're making them," Pennisi would say in meetings. "You may not like them all; all I can ask is that you respect them."

To embed the concept of "commerciality with a heart" into the organization, Pennisi set up a number of new initiatives that wouldn't have been feasible before. On the business side, there is now a fund available to private pensioners and a new insurance company that serves the existing public-sector-employee base. There is also a new initiative to counter domestic violence. "We have a partnership with a fantastic non-profit, which runs one of the busiest help lines in Australia.

We financially backed one of their counselor positions, which means about 4,000 more people a year can get their calls for help answered." The company is also working with Australia's indigenous community. "Taking a stand in this way is important because it helps attract and retain the best people, but also because it gives us moral stature," says Pennisi.

His role as a Wise Advocate in the organization is paramount in his mind. "All leaders cast a shadow," he says. "It's really easy to be a star when times are good. But when things are difficult, and particularly during times of great change, you learn what leaders are made of. People's radars are set on ultra-sensitive, and they're looking at every move the executive makes, and every word that's said. They want confidence that you know where this is going. Then they'll follow you."

It's worth emphasizing the importance of the Wise Advocate's third-person perspective on first-person experience in this story. QSuper's initiatives are a departure from its past and could be seen as ephemeral or quixotic. While bringing them to fruition, it's important to keep an eye on the company's core goals and how they contribute. This means seeing the company, and (for Pennisi) seeing his own thinking, as an external person might see it—once again, calling on the perspective of the impartial spectator or Wise Advocate. For many business leaders, including Pennisi, an executive coach would provide that perspective at first. But over time, it becomes internalized; the Wise Advocate of the mind is habituated. Coaching or other forms of mentoring also adjust, as the individual leader graduates to more and more advanced questions about the aspirations and responsibilities of the enterprise.

Shell Oil: Leading Through Messiness

Another example is drawn from the story of Shell Oil in the 1990s. Based in Houston, this was the U.S. enterprise within the Royal Dutch/ Shell global group of oil companies. It was the largest and most profitable company within Shell, but it was also facing a time of severe cutbacks, dating back to the collapse of the oil price in mid-1980s. After 1991, which saw the worst financial performance in the company's one-hundred-year history, Shell Oil put itself through a major restructuring, including a high number of layoffs. But annual performance did not improve. A new CEO, Philip E. (Phil) Carroll, was promoted from a role running the company's administrative services. He and the board then resolved to solve the problem by reorganizing the company based on Charles Handy's idea of "subsidiarity": moving decision making to the most decentralized level possible, while maintaining close coordination among all the parts of the company. Some parts of the company would have their own P&Ls for the first time. It would also mean moving away from the authoritarian culture that had been prevalent at Shell Oil and getting people to be candid about the business—and the human dynamics that shaped the business—in ways they hadn't been before.[3]

A striking feature of the Shell Oil story was the number of people who put themselves on the line for the transformation: changing old habits, rethinking practices, and consciously trying to foster a new culture that would encourage people to make a genuine commitment to the enterprise. One individual who stood out, even in this context, as a Wise Advocate within the company was Linda B. Pierce. She had been a quality-improvement manager in the information-services division, became an internal coach and organizational-learning specialist, and then rose to a position as liaison between the transformation effort and the leadership council.

If Pierce had held to a Low Ground perspective, trying to please everyone she worked with, the administrative role could have become a thankless transactional job. But Pierce made it strategic. She was visibly conscious of the power of habitual thinking—her own and others'—with regard to the changes Shell was able to make. She repeatedly reminded people of the stakes in this game: *This transformational effort is a rare opportunity to make this organization work effectively.* She was hard-nosed and pragmatic: she was not afraid to give people a hard time, including the company's top executives on its ten-person leadership council, if they didn't keep their promises. But she did it with a clear and evident respect for everyone in the game, including a few of the top executives who were not trusted by others in the company because of promises they had broken in the past. She became the kind of role model for integrity that everyone looked up to. There were people who outdid themselves in training others, or managing performance, or raising questions, simply because they didn't want to let her down.

Toward the end of her time at Shell, Pierce published a short essay on the role she had played: "A Personal Checklist for Leaders Facing 'Messy' Experiences." She wrote that when she was stuck, she would check the following principles. "If I can't answer these questions affirmatively, that tells me that I need to get myself centered again."

- Do I have constancy of purpose? The phrase comes from Dr. W. Edwards Deming. I am only effective if I understand, in my own way, why we are investing our efforts.
- Am I patient enough? I expect too much if I expect the system to become perfect—today. I can become angry when it doesn't match those expectations, and I undermine my own ability to contribute.
- Am I impatient enough? Do I stay clear of complacency, or collusion with the status quo?
- Can I muster the necessary courage? As a change agent, I must give voice to help others hold the mirror to see the warts, gaps, and problems. It's sometimes difficult.
- Can I keep an optimistic bias? "Change agents" who perceive the system as too biased and corrupt will not be able to do good work in corporations.
- Am I opportunistic? Since change is mostly an emergent process, I must be alert to unplanned, but well-prepared interventions.
- Am I approaching people with compassion? From time to time, I can observe others displaying defensive or bizarre behavior. I have learned that in just about every case, people are being as good as they can be at that moment.
- Do I conduct my work with love? When I love people, I want to help them be all that they can be. If I can bring that attitude to my work, then I can be effective.[4]

Note two things about this passage. First, Pierce is focusing on her own behavior; she isn't judging others. Yet by holding herself visibly to this standard, she became a role model for her colleagues, who regularly asked her for advice and guidance throughout the Shell Oil system.

Second, she is repeatedly putting herself in the position of the Wise Advocate. Why are we investing our efforts? Are we patient enough or too patient? She speaks of holding up a mirror, being alert, observing others displaying defensive behavior without taking it personally. She is mentalizing ("They are being as good as they can be at that moment") and mindful ("If I can bring that attitude to my work, then I can be effective"). She has put it all together, seamlessly, in a habitual practice that is both deliberate and natural to her.

Revaluing Is Step Four

These stories make it clear that Pennisi and Pierce were highly skilled at the first three steps. They knew how to relabel deceptive messages; how to reframe new ideas; and how to refocus the efforts of the enterprise to build new habits of mind.

And then they went further. They embodied strategic leadership in ways that were sometimes easy to see and sometimes so subtle that people didn't realize their impact until later. In everything they did, they fostered an organizational context—practices, processes, and cultural efforts—that brought people to the High Ground and made it a way of life.

For example, Pennisi regularly conducts question-and-answer sessions with employees, usually with about two hundred people at each. After a minute's conversation, he sits on a stool and answers questions. Typical of the questions is one he received at the first session he conducted. He talked about the decentralized decision-making system and the commitment it would require from each of them. Someone put a hand up and said, "That's really good, Michael. But when are you leaving? How long do you expect to stay here as CEO?"

Pennisi had a ready answer. He said he believed that a CEO's tenure should last between four and seven years. After that, it was time to hand over to the next person. "This was an important question for me," he later recalled, "because it was so personal. Because I was willing to answer [that question], people realized that they could ask me anything and I would answer it, and there would be no repercussions for the questioner."

At Shell Oil, Linda Pierce set up guidelines for in-depth coaching and training that were implemented at every level. For example, there was a new practice in giving direction: leaders could no longer give commands without explaining their rationale and their connection to the company's strategy. Just listening to and considering these types of explanations would move people to High Ground mental activity: *What are the speakers thinking? What are they going to do next?*

Other leaders we have written about in this book have made similar efforts to institutionalize High Ground activity, and thus revalue their leadership. Bill O'Rourke at Alcoa was doing that when he focused on safety as a vehicle to change thinking in aluminum plants and (most importantly) explained his rationale for doing so. Reid Hoffman encoded his reframing of business relationships—one step beyond the transactional—in his design of the LinkedIn social network. At QIC, Marcus Simpson explicitly recruited people who could contribute outside perspectives to his team:

> We deliberately assembled a team of people with different personalities. We've figured out ways to work together and to have tough conversations. . . . This balance is really useful, because when I have a wild idea I'll receive a wide range of views. It has been frustrating at times, but I believe it has produced better outcomes for our clients.[5]

Relabeling, reframing, and refocusing are prerequisites for revaluing. They lead up to it. They help you develop the necessary habits of mental activity: thinking about your thinking (metacognition) and paying attention to where your attention is focused (meta-attention). And they give you practice in moving your mental activity from Low to High Ground, rewiring your brain in the process to make it easier.

If you are wondering whether the patterns of mental activity from the previous three steps have genuinely affected the biology of the brain, your level of success in revaluing will tend to tell you. You become aware of changes in your thoughts, feelings, and intentions. As you grow accustomed to your new proficiency, you gain confidence. You recognize your strength as a leader; the necessary thoughts and attitudes are now embedded in your brain circuitry.

It's like learning to play the piano. When you start, you have to pay a great deal of attention to your fingers, training them to be skilled. Once your fingers are trained, you're in a more proficient state—self-directed neuroplasticity has occurred. Now you can move on to think about your playing at a deeper level. It's the same piece of music, but you think about tempo, variations, and the performance. You watch yourself at the piano. How can you connect with this audience? What might you do differently? What will be the same? As you gain in skill, you connect more deeply with the source of skill within yourself that gives your playing its character and quality. You aren't complacent; you know that you can always improve. But you recognize the value of the mastery you have achieved.

Revaluing and the Core Group

In the revaluing step, you will find yourself mentalizing about the organization in the same way you considered your own identity in the previous steps. *What is this organization about? What is it trying to do? What is it likely to do next? What would life look like if it were more successful at realizing its highest aspirations?* This is a different set of questions from the Low Ground concerns: *What do people want, and how do we please them?*

To answer these questions, you need to understand the concept of core group dynamics. The core group is our name for the people in any organization who are considered important by everyone else. The organization will act only on their perceived behalf, to fulfill their perceived needs and priorities. This is not because they are authoritarian or because people are following them blindly. It's because of the way people have to make decisions in a complex, fast-paced world.

Consider the thinking behind many of the everyday decisions that you, as a leader in the middle of the hierarchy, must make in your organization. The decision is complex, and no rule or guideline could cover all the possibilities, but there isn't time to think it through in detail. After all, you've got other decisions close on its heels. The most expedient thing you can do is to make your decision in light of a mental construct of the few people you think are relevant: *How does this fit with Frank's plan? What will Sally want me to do?* And, most tellingly, *I don't want to be the one to walk into Joe's office and tell him it's not going to happen.*

This is a Low Ground approach, to be sure; but even on the High Ground, even consulting your Wise Advocate, you need to consider the core group. These individuals stand in for the organization as a whole, which would otherwise be too complex to think about. They are not necessarily the people at the top of the hierarchy (though the CEO, some members of the top team, and the head of your own division will probably be included.) The criteria for membership are informal and may vary from one company to the next, or even from one day to the next, depending on how stable the culture of the enterprise is. It may include some local bosses, some key functional leaders, perhaps the head of the labor union or of a critical subsidiary, or someone who serves as an unofficial conscience—a recognized Wise Advocate, regarded as the voice of integrity. Different parts of a large organization often have different, albeit overlapping, core group members. Thus, the head of learning and development or the chief procurement officer is in the core group for everyone within their respective functions, but not so much for people in the business units. Since every company has its own core-group dynamic, it's difficult to generalize; in one company, the glass ceiling may keep women and people of color from rising to the top, while it may keep white men over fifty from rising to the top in another. No matter who they are, the core group members all have one thing in common: legitimacy in the eyes of everyone else. They stand, in people's minds, for the priorities of the enterprise.

Not all decisions are based on the needs or priorities of Frank, Sally or Joe—but enough decisions are that it determines what the organization is capable of doing. Moreover, these decisions are not based on reality, but on the *perceived* needs of Frank, Sally, and Joe. Like deceptive organizational messages, the perception of the core group's wants and needs is based on incomplete information, including the way they have communicated in the past. As a symbolic construct, the core group is like the inner circle in a family—the people on whose perceived behalf decisions are made by everyone else. (The thought *I don't want to be the one to tell Grandma the wedding's been called off* is close in tenor to the thought *I don't want to be the one to tell Frank that we let the deal fall through.*)

Like a family or community, the organization as a whole will do only what is perceived to be in the core group's interest. The hierarchy may dictate who has authority, but the core group determines who has legitimacy. Resistance to change occurs not because people fear change, but because they fear the consequences of contradicting the perceived priorities of the core group. *What if I go out on a limb and then Harry changes his mind about the strategy?* They know that if Harry reframes a particular point of view in a particular way, many others will do the same.

This core-group concept may seem harsh, but it reflects, in our view, the simple nature of collective humanity. Life is simply too short to think through every decision completely. People at work need guidance, and if it's not going to come from a Wise Advocate within, it will have to come from somewhere else. Understanding the core group is important in the revaluing step because it shows you how the enterprise is limited, and it demonstrates where you, as a strategic leader committed to the organization, will have to focus your attention.[6]

But the flexible nature of core group dynamics is one factor that gives you an edge with revaluing. Every time people make decisions, even subconsciously, that support an individual in the core group, they strengthen that person's position. Like brain circuits, these organizational circuits get strengthened the more they are used. By finding those core group members who represent the right values for your enterprise, and speaking to that value as part of your own leadership presence, you can help shift the organization in a more constructive and valuable direction.

When you revalue, your task is not to overthrow or eliminate the core group, or even to protest against it. You certainly don't want to talk about its influence openly in a negative fashion; that would undermine your own efforts. Your goal is to align the values of the Wise Advocate, and the organization's true self, with the core group's priorities. You do this in the same way Wise Advocates have always operated: by relabeling messages in a nonthreatening way, reframing a new narrative that can help the organization succeed, and refocusing the attention of the enterprise around those concepts.

This may seem like a daunting challenge. In most organizations, the core group is associated with the Low Ground. There is a natural affinity between the Low Ground and anything expedient, and chances are the core group's priorities revolve around solving problems as quickly and broadly as possible.

But when you mentalize about the core group—when you start thinking about what the core group members are thinking and why, and what they are likely to do next—you will tend to see aspects of their strategy that weren't obvious before. You start to balance the core group's immediate and obvious messages with inquiry that goes deeper into mindfulness and mentalizing. *Who is the core group? What are they thinking? What are they likely to do next? How do I relate to them?* This attentiveness gives you qualities that some core-group members will probably recognize—among them, that you are thinking not just of your position but of the whole enterprise. It will also help you build connections with others who feel the same way you do about the organization and its priorities. When people throughout the enterprise start paying attention to High Ground issues and achieving results through them, that will change the core group dynamic as well, because leaders in the core group owe their legitimacy to the willingness of others to follow them.

An organization needs people who remain loyal to the core group but who don't accept its pronouncements at face value. If your mental activity is coming from the High Ground, you may be well-equipped to talk to members of the core group directly about their priorities. (Few people do. Rather than ask core group members what they are thinking about, many people guess what they want, because it looks more professional to know and provide what the core group seeks. Unfortunately, these guesses are often wrong, erring on the side of expedience. In many cases, core-group members appreciate the candor and courage it takes to find out what is truly important to the enterprise. To be sure, you recognize the power dynamics. There may be things you don't say out loud, but you have a clearer idea of why those limits exist and how to make the most of your position—assuming that you care enough.

A Wise Advocate in the Core Group

What if you, yourself, are in the core group? How then do you engage the High Ground? The legitimacy you have, as a core group member, will move you and your organization more rapidly in a good direction. But there are also pitfalls to watch out for.

As with the mentalizer's paradox, the experience of being in the core group of a large organization makes it harder to stay on the High Ground. People throughout the enterprise pay close attention to everything you say and do, and especially to where you pay attention. They make it their mission to guess what will please you, and to try to deliver it. It's extremely difficult to avoid the feeling that you are the center of things, the most important person in the room. And that diminishes your effectiveness as a High Ground leader.

Consider the story of Noel Lord, state sales manager for the Macquarie Bank in Queensland, Australia. This position is responsible for sales and customer services for financial-planning clients. At the time this story began, a few years ago, he oversaw a pyramid of close to a hundred people, with a team of nine direct reports. He became familiar with the four steps when he asked an executive coach to help him recognize the gap between his managerial intent and the impact of his actions.

Lord was very much in the core group of the enterprise. But he was also a problematic manager. "I was ignorant and unworried about the impact of my communications style: for example, I would hold others accountable in inappropriate ways, warts and all."[7] His intent was to help people perform better. But few people could see his intent; it was often below the surface.

For example, one of his salespeople averaged only three or four appointments a week instead of the customary ten or twelve, and Lord—with the intent only of emphasizing accountability—took him to task at a weekly staff meeting. "Are you kidding me?" he recalls saying. "This is the third week in a row." Then he turned to the rest of the team: "Are you telling me you think this is acceptable?" Behavior like that from a core-group member incites the Warning Center and reduces productivity, not just for the individual being targeted but for everyone in the room. They come to expect the boss to turn on them. The deceptive message usually takes this form: *This company views everyone who works here as replaceable. No matter what we try, we will be blamed and bullied.*

After a while, the group's performance could no longer be ignored. After talking with his coach, Lord sought feedback from a few people who could speak honestly and openly to him about his management style and the way people reacted to it. It was eye-opening, and it led to a pivot point—a moment of relabeling. For the first time, he saw himself as if from his staff's eyes, and recognized his behavior as a source of intimidation. "I apologized to him and the team separately, admitting that it was inappropriate behavior." Lord added, "If you don't have your appointments sorted out before a meeting, come and see me first."

The opportunity for refocusing came a few days later, in another team meeting. This time, the salesman was aggressive before the full group, blaming Lord for the lack of appointments as well as for the inappropriate comments. "I hadn't been respectful of him in front of his peers, so I could not expect his respect now," Lord said. "But we agreed to have one-to-one discussions ahead of each meeting, as well as team accountability—not individual blame—during the meetings." By taking on the responsibility of meeting with this individual every week, Lord made them both accountable for his performance.

The need for higher sales volume stayed constant, but Lord had now sent a message to the salesman—and everyone else on the team— that they were valuable enough to have their progress be worth the boss's time. This was reinforced, week after week. People noticed not just what Lord said, but that he was now paying attention to them in a new way. They built a new relationship with him, and began revaluing their relationship to the company. At this point, as a key member of the core group, he was setting an example that was influential throughout the company.

Could he have done all this in his previous Low Ground frame of mind? Perhaps—but he would not have had either the rationale or the energy to carry it through. There would have been too many distractions, carried by the deceptive organizational messages running rampant through the system.

When you are a core-group member, you may often find that you set off reactions unintentionally, sometimes through emotional contagion. Your ideas, thoughts, and even moods will spread to those around you, simply because of your core-group status. It can happen even to leaders who are known for their equanimity and enthusiastic support of the people they work with. One such leader brought a group to a restaurant to celebrate an individual's promotion. But in the middle of the dinner, he got a call from someone with a complaint he had been dreading, on a matter he couldn't fix. This changed the leader's mood. When he hung up, he was obviously brooding about it. Another person at the table asked, "Is there anything I can do?" The leader snapped back, abruptly, "No!" Everyone else at the table shut down; the mood of celebration faded.

Conversely, when a leader makes an effort to be broadly respectful, it ripples through the organization positively. In one company, people in marketing and R&D were used to carping at each other. Then one of the R&D leaders explicitly said, "From now on, I am not going to blame anyone in marketing without checking out the story with them first." That comment, in itself, percolated through both functions and had a rapid effect in improving the atmosphere.

Even on the Low Ground, emotional contagion matters; people who display overt anxiety do not succeed. But the "straight from the gut" transactional approach of the Low Ground is enormously contagious, especially from members of the core group. People assume that, like the core-group members, they need to play a game of winners and losers in which leverage comes from giving people what they need. The High Ground, however, is just as contagious. When the core group seeks a larger strategy, others recognize this and begin to play the High Ground game themselves. If the core group is seen to make the kind of moves a Wise Advocate would support, the rest of the organization will do the same. That's why we believe that behind every great organization is a core group of strategic leaders.

Leading Outside the Core Group

What if you're not in the core group? Does that mean you won't have influence or be treated fairly? Experience and observation suggest that the opposite is true; if you know how to cultivate the inner voice of strategic leadership, you can accomplish quite a lot—for yourself and your enterprise—from a position outside the core group.

Your role in the enterprise will always be transactional, in the sense that you are paid a salary and benefits only in exchange for the work you provide. You hold no equity; your position carries no official weight beyond its limited parameters. But you can still give voice to the organization's most important needs and priorities, in a way which generates respect and influence. In other words, you can still revalue yourself as a leader, observing yourself in action at the same time that you act.

This type of revaluing is best accomplished in collaboration. You step out into the enterprise, finding allies, advisers, and cocreators of a new reality. Some of them may be top leaders of your organization; others will be people who feel some reason to make a strong commitment to the enterprise's success. The one thing they have in common is that they care: like you, they feel a strong stake in the outcome. Each of them, in his or her own way, is going through a similar process: recognizing that some of the organization's messages are deceptive, reframing them, and fostering new habits of thought.

The conversations you have with these colleagues will shift away from Low Ground concerns about status and office politics: *What do they want? Who is winning?* They'll move instead toward High Ground concerns: *What are the leaders thinking? What will they do next? Why do they think this way? What does the organization as a whole think? What will it do next? And what does it need to do next to really fulfill its aspirations?*

When you persevere on the High Ground and enter the Higher Ground, especially when abetted by mindfulness practice, you will notice a change in the way people respond to you. They are picking up on the Wise Advocate in your words and approach and passing it on to others in the organization. The more consciously you do this, the more adaptive and accelerated the change. Here's a story from an executive in a financial services firm in the United Kingdom that illustrates the point:

> I was at a meeting in our organization talking about the aftermath of the Brexit referendum. The vote to leave the European Union had raised a lot of business issues for us. Our revenue volumes were likely to go way down, unless we took immediate measures—and there was a lot of disagreement about what measures to take. The conversation grew more and more transactional; people heatedly debated the points as if their own jobs would be on the line if they lost. The senior-most person in the room, one of the top executives of the company, did something smart. He said, "I know we've got a lot to cover, but I'd just like to think for a minute. Let's stop."
>
> There was a beat or two of silence, and then one of the most argumentative people in the company, somebody who always has a lot to say, began to make a case for something. And everybody just looked at the arguer.
>
> Then everybody stopped.
>
> Five minutes went by before we started talking again. We were talking more slowly; taking more time to breathe. And we came to a solution that was much, much better. It felt like we had actually tapped the potential of everyone in the room, instead of just one or two fast talkers. It was like seeing the High Ground come to life.[8]

We've seen many sessions of this sort that bring the Wise Advocate to the surface. These sessions are sometimes conducted with a high level of artistry; the invitations, the environment of the room, and the opening conversations are carefully designed. At other times, they occur spontaneously. The critical factor is for the Wise Advocate to step forward, drawn by a genuine wish to move the organization closer to the Higher Ground and having enough experience with the four steps to be capable of making it happen.

The High Ground and Higher Ground can be invoked by certain types of session designs: a thoughtful scenario planning exercise; a Ladder of Inference exercise; a powerful executive-coaching interaction; a dialogue; an exercise in mutual understanding among former foes, like the Truth and Reconciliation commissions in South Africa and Colombia; a really good competitive or customer analysis, where normal business practices are revamped to become more like mentalizing: *What are our competitors thinking? What are our customers likely to do next?*

In all these cases, and others, it's tempting to think that the process is the critical factor—that the Wise Advocate has been brought to life through the design of the exercise. But everything depends on the way it's handled. We have seen sessions that might have yielded a great leap forward, but that instead put people put on the spot or drowned in trivialities. In the history of management, there have been many efforts to develop processes for motivating people or galvanizing performance.[9] The sessions that work are successful because there is a relatively self-aware leader in place, navigating the tensions, retaining their loyalty to the organization while keeping their self-respect (and their respect for everyone else) in place. Revaluing develops that type of leadership.

For revaluing to work, you need to maintain a perspective unfettered by your bias; you have to look at reality clearly and dispassionately. You also need a high level of commitment; you have to care—about yourself, about the people around you, and about your organization as well. (If you don't care enough about the organization to make that sort of commitment, then a revaluing process like this may raise your awareness to the point where you make the choice to leave.) They do not reflect the needs of expediency: that's the Low Ground. Your values are reflected in the commitment you make to your organization's potential future; to helping it become the kind of enterprise you would be deeply proud to belong to. As we said, this kind of commitment cannot be tackled alone; nor can it be tackled simply through the expression of ideas. It requires actions: pilot experiments or new ventures that can test your ideas about what the organization should do. Successful leaders try things, not perfectly at first, but continually increasing their competence, reach, and the humility that comes with understanding the enormity, and the importance, of the task at hand.

You will probably find that the revaluing step raises questions about your own aspirations, and the sources of your idealism. In the mid-2010s, one of the authors of this book (Josie Thomson) emerged from a near-fatal bout with cancer determined to "save the world":

> I developed a training that was about how to have it all—a brilliant professional career, the noblest aspirations, and a great private life—without going mad or getting sick. Although my cancer experience was the catalyst, I didn't mention that aspect of it. But the harder I pushed the idea, the more my [executive coach] business declined.
>
> I went back to the Wise Advocate in my mind. "Okay, I'm here in service. What would you have me do?" The reply that came to mind was: "Remember who you are."
>
> I wrote one email with that idea in mind, and within twenty-four hours I had ten coaching clients paying me $25,000 each. In the email, I simply said, I'm only taking on twelve clients in the next twelve months. I've learned how to manage extraordinary stress in the past few years, in part through my experience with cancer. You can learn it too.[10]

We don't suggest abandoning your ideals. We suggest tempering them by using the Wise Advocate's perspective to continually focus your attention on the value you bring, and how to increase that value.

The Leader's Commitment

The higher you rise in your organization, the more intensely you may feel the pressure of the Low Ground. To think about the way this can affect an organization and a brilliant strategic leader, consider the story of someone who squandered an immense amount of fortune, talent, and reputation. He destroyed one of the great business empires of the twentieth century because he could not get in touch with his own Wise Advocate.

Howard Hughes, the chief executive of the Hughes Tool Company, Trans World Airlines, and the RKO motion picture studio, was an exceptional man. He inspired devotion and admiration among employees and the world at large. He was a champion aviator, one of America's earliest billionaires, and an entrepreneurial leader who pushed three industries to transcend their limitations: motion pictures (with the epic airborne special effects he pioneered), aviation technology (with the world's fastest and largest airplanes), and airline transportation (with cross-continental flights above the clouds).

He was also famously stricken with obsessive-compulsive disorder in a way that destroyed both his life and his companies. A variety of self-destructive habits governed his life; he knew they set him apart, but he often couldn't tell whether they were right or wrong. Many of them, like his habit of repeating sentences, were obvious to his subordinates, and to himself; but he couldn't control them. Some of them involved deceptive messages: his misperceptions of value, in the form of perfectionism and "all-or-nothing thinking," led to a profound lack of confidence in anything he couldn't fully control. "All of Howard's delays [as producer on the film] *The Outlaw*," wrote his biographers Peter Harry Brown and Pat H. Broeske, "his indecision over the construction of the flying boat, his thousands of takeoffs and landings [in test flights], were all merely symptoms of an illness."[11] In the mid-1940s, for example, he became so afraid of eavesdroppers and spies that he conducted meetings in cars or bathrooms, and so afraid of accidentally contracting illness that he required his aides to put on white gloves before taking an object to him.[12] At the same time, he was fearlessly convinced of his own exceptionalism, that the ordinary rules did not apply to him, and that his perceptions, feelings, and thoughts were the unchallengeable truth.

Hughes recognized these problems himself, especially in his early years, but he had an additional factor to deal with that most sufferers of OCD do not have: his destructive mental habits and compulsive behaviors were reinforced by leadership contagion in the organizations he led. At times, for example, the companies took on his own perfectionism and delayed their product launches, even when he didn't make that decision. At other times, when Hughes mysteriously disappeared—sometimes into long flights, other times into hotel rooms, and once into a seven-month secret nervous breakdown—the leaders of the companies had to manage on their own, never knowing when or how he would return to reengage. So they fought with and constantly undermined each other.[13]

The effect on his companies was to give them their own self-destructive story lines. Hughes Aircraft was notorious for poor management, but few tried to improve it; "Howard would just put everything back the way it was," they said. They also felt that fixing the company's management style didn't matter, because another deceptive message held sway: If they only got the aircraft design right, the world would beat a path to their door.

Had Hughes taken the time and given the attention needed to see himself as others truly saw him, his business legacy might have been very different, untarnished by the years he spent as a bitter recluse in a hotel room, and by the destruction of much of the value of his enterprises. Hughes Aircraft, for example, lost most of the business it could have gained during the ramp-up of military technology enterprises in World War II, in part because Hughes would not allow anyone else to oversee completion of its innovative "flying boat" or resin-based bomber-plane projects (the D-2 and XF-11) during his long absences.[14] By the time he reached his fifties, all his contact with the outside world was gated through handlers, who swindled him out of millions of dollars. This ultimately forced him to sell Hughes Tool, the oil-drilling-equipment company he had inherited and that was the primary source of his fortune.

His story is an extreme case, but many other leaders get caught in similar dynamics. Indeed, if you are a leader with any core group presence or an influential persona, your own impulses and inclinations may sometimes mislead you and draw you away from the High Ground. It begins the way Hughes' story began: as a person of influence, you become caught up in the experience of your own success. Every habit reinforces feelings of self-importance; the practice of leadership becomes a form of narcissism, in which your own status and emotional state take on disproportionate significance and you lose sight of everything else. You no longer mentalize because there is no need to think about what other people are thinking or what they might do next. You are not mindful, because you must press on to clear the next hurdle, and the next, and the next. You probably have too much to do, but you don't find it easy to delegate, in part because delegation takes a more focused, slower form of attention than just doing it yourself; and in part because no one else around you seems as capable as you are.

Meanwhile, everyone around you feeds your desire for praise (and praiseworthiness), and few are willing to give you the perspective you need. If they do, you tend not to listen to them. You may even fire them for contradicting your views, as Hughes did with several of his most capable subordinate leaders.

More likely than not, you are still creating deceptive messages. You may even be more skillful at creating them; your messages may be more seductive, more idealistic, and yet more dangerous than they used to be. *We need to lead from the gut. We're so successful that no one can stop us. If they try to hurt us, we'll hurt them harder. We don't need to change anymore; we've arrived. We've paid our dues.*

How can you overcome this trap? How can you transcend the limits of your own narcissism, especially when everything around you reinforces the message of your own self-importance? How can you escape your own blind spots? How can you change your ways when there is no clear compelling reason to do so, and when it isn't even obvious what will happen next if you change?

The only answer we know of is the Wise Advocate within yourself: the impartial spectator that Adam Smith discerned, plus a genuine commitment to your own true self. One source of hope in the world today is that people of influence, at all levels of society, recognize this. They are cultivating their Wise Advocates. They are mentalizing—thinking in depth about what other people are likely to think and do. They are improving their executive function—developing stronger cognitive flexibility, working memory, and inhibition control. They are practicing applied mindfulness—paying attention to their own thoughts and actions, with a clear-minded rational perspective. In the process, they are improving their own capacities for leadership.

Calling on the Wise Advocate often flies in the face of many of the things that leaders take for granted: that power makes you feel good; that expedience works; that decisions should be made for the right reasons; that you need to do more, more, and more. It also means going against the persistent influence of those around you, which is remarkably hard to do. Social psychologists like Naomi Eisenberger have documented that the feeling of self-esteem is determined, at least in part, by social cues.[15] Approval from people around you leads to the kinds of intense feelings that people associate with morality.

But moral philosophers, including those as pragmatic as Adam Smith, recognize that social cues are often deceptive. *How will other people feel about this?* is a Low Ground question. To be a genuinely strategic leader, you need to recognize when you are on the right track, even if no one else agrees. On the High Ground, self-esteem is not determined by whether other people like you, or even by whether you're praiseworthy; it's determined by your sense of how you would be seen and judged by people you respect. Self-esteem depends on the internal values epitomized by your own Wise Advocate, in touch with a purpose larger than yourself, even if that purpose is not fully articulated.

Whether religious or not, anyone can recognize the meaning in the Bible's question "What does it profit a man to gain the whole world and forfeit his soul?" In the research on mental activity that we described in chapter 3, Shihui Han found that the devout people he studied had internalized this question, to the extent that it deeply affected their thinking patterns and brain responses, and that this deepened the same cycles that were inherent in advanced leadership. Those who believed deeply in a spiritual path—in this case, that Jesus was their model—when thinking about a reward, did not jump to "What will it cost me to get that?" Instead, they jumped to "What would the Wise Advocate say about the value of that reward?" And their brains were activated accordingly.

This finding is consistent with many others. It helps explain why contemplative traditions have been so influential in human history, and why religion has been so important to communities. It also explains why the practice of mindfulness is popular in business circles; it brings to bear a way of thinking that is often missing in daily life.

We are not saying that you have to be devoutly religious to gain the benefits of being an effective leader. But we are saying, with a high level of confidence in the experimental evidence, that the frame of mind evident in the devout Christians studied by Han is akin to the frame of mind evident in highly effective strategic leaders, and even more evident among transformational leaders. Whatever your personal religious belief (and the three authors of this book have three very different sets of personal beliefs), it is clear that your capacity for effective leadership is closely related to your own values and understanding. You cannot be a great leader unless you are plugged into something larger than yourself, something that helps you move past the Low Ground. History suggests that you may not be able to get there on your own.

This creates a major dilemma for organizations, particularly secular business organizations. We want High Ground leaders. But how can a business require its leaders, or its employees for that matter, to strive to transcend their human limits? Is it fair to ask that they do so? How can we make a very personal commitment part of the requirement for worldly success? Is it fair to require people to transcend their current moral limits, even if the requirements of the job call for them to do so? And if we require no less, what is to prevent the organization from falling into a different sort of Low Ground, in which people are coerced in the name of the High Ground itself?

These dilemmas have been recognized for centuries. For example, the seventeenth-century scientist and mathematician Blaise Pascal wrote that it isn't possible to stay on what we call the High Ground without divine intervention. About 100 years later, as we noted in Chapter 3, Adam Smith articulated the same dilemma in different terms: there is no way to effectively impose morality from an outside structure, neither one grounded in laws nor in community norms and attitudes. Morality must come from within.

The challenge of finding morality in a world as complex as ours, may simply be too difficult. But what other choice, for a strategic or transformational leader, is there?

Songwriter and cancer survivor Darlene Zschech, who is known for her presence in the Christian spirituality movement, once said, after her recovery, that—though she would never want to go through the experience again—there was one thing she missed. "That's when I thought it was over and heard God say to me, 'Now breathe in; now breathe out. Now breathe in; now breathe out.'"[16] Inherent in the Wise Advocate is seeking an experience of that sort—not being able to command it, but being able to hear it, perceive the value in it, and draw upon it. This is not the same as thinking you're the center of the world; it's closer to thinking that you are fortunate to be in this spot, to have these resources, and to be able to make the most of them—because you are tapped into something larger than yourself.

It doesn't make rational sense. Neither do the most effective leaders. Your ability to be an effective strategic or transformational leader will demand of you an ability to transcend the limits of rationality—in a way that you choose. We understand that the available answers on how to do this are not satisfactory to many—and may be intrusive in an organizational context. But The questions can't be avoided: *Are our goals, as an organization, the right goals? If we are maximizing profits, what are we maximizing profits for? As an individual, have I chosen the right goals for myself? Am I approaching them in a way that I will be glad about when I look back at this episode later?*

All three of us have wrestled with questions like these directly. Among us, we have written about narcissists, obsessive-compulsive individuals, people who have attempted suicide, organizational heretics, cancer survivors, and mysticism. Our efforts to develop an understanding of this have taken Jeff to a deep exploration of Christian values and meditative traditions;[17] Josie to an in-depth inquiry into the nature of personal growth; and Art into extensive research into group dynamics and the insights underlying it.[18]

In the end, we come back to the reality that it is intrinsically better to move your mental activity to the High Ground. This doesn't happen just once or twice; it happens every time you have a decision to make. Back in Chapter 1, we summed up the core message of this book in one sentence. Here it is again: *The focus of your attention determines what happens in the mind and the brain during these critical moments of choice, and determines what kind of a leader you can be.* Over time, your mind will develop distinctive neural patterns, subtly different from those of other people, because of this ongoing practice.

To us, this is a message of hope. As a strategic leader, simply by being drawn into a position of influence, you've accepted a commitment. You are the only person who can choose how to exercise that commitment. With the right habits of mind, you can continually call on your inner judgment enabled by your Wise Advocate. Working with the right other people, you can continually move your organization to a better place. Your humble efficacy, when channeled in this way, becomes a source of power that can change the world.

You may or may not deserve this role. You may or may not be rewarded or recognized appropriately. But here you are. You've received a profoundly valuable gift. You're a leader on your way to the High and Higher Ground.

Acknowledgments

FIRST AND FOREMOST, we thank you, the readers. If this book intrigues you and you want more contact with us, please visit the associated website: WiseAdvoc8.com.

Many people provided the context that allowed this book to come together. We are profoundly grateful to all of them. There are a few who we wish to thank or credit by name:

Rebecca Gladding: Her work with Jeff Schwartz on *You Are Not Your Brain* laid the groundwork for some of the ideas herein. She also introduced Jeff to Darlene Ricker, who suggested the term "Wise Advocate," for which we are forever grateful.

Faith Florer: She provided extremely valuable commentary on the manuscript and did indispensable work in compiling the first draft of references.

David Rock: He introduced the three of us to each other and—with Al Ringleb and Lisa Rock—created the Neuroleadership Summits, in which we participated in the early years. David and Lisa Rock have also been pivotal to Josie in every aspect of her personal and professional life as an executive coach, and gracious colleagues to Art.

Melanie Rodier and Jonathan Trippett facilitated the video production that demonstrated that we had something to say together. Thanks also to the team at strategy+business, which published several articles on themes related to the book.

At Columbia University Press, we were privileged to work with Myles Thompson (who, as editor-in-chief, saw the value in this book and its authors from the start), Stephen Wesley and Ryan Groendyk (editors who got us started), Brian Smith (the editor who brought us to the finish line), Meredith Howard (promotion), Ben Kolstad (a consummate managing editor at Cenveo Publisher Services), and Marielle Poss (the press's EDP director) on the development of the manuscript. Information graphics specialist Linda Eckstein produced the three elegant and critically important illustrations in this book.

Some individual conversations were important for us. They include conversations with (among others) Stephanie West Allen, Drake Baer, Brittany Beall, Graham Bronczyk, Knox Brown, Angela Duckworth, Emily Falk, Pablo Gaito, Ann Graham, Heidi Grant, Tiffany Gray, Sean Illes, Iliyan Ivanov, Jon Katzenbach, Annette Kramer, David Lancefield, Doug Lennick, Maud Lindley, Susan Mackie, Ed Martin, Mary Beth McEuen, Kathy Meehan, Martin Monti, Jim O'Toole, Carolyn Parkinson, Juliette Powell, Wendy Quan, Susan Rabiner, J.M. Reid, John Rives, George Roth, David Sable, and Malcolm Thompson.

We wish to thank Josie Thomson's executive-coaching clients for sharing their stories—those who appear by name in the book and those whose names are withheld.

Jeff Schwartz would like to personally thank Peter Whybrow, MD, for his many years of loyal support at UCLA, and Stephanie West Allen for her invaluable help in keeping abreast of internet coverage of neuroscience-related topics.

Josie Thomson would like to personally acknowledge Dan Radecki and Josh Davis for inspiring, supporting and encouraging her to complete a master's studies in neuroscience, despite her own encounter with a brain tumor while undertaking the studies. She would also like to acknowledge her children Kristen and Harrison—who were supremely patient and supportive during this project while they were completing their own studies—and the many late nights shared at the kitchen table studying and working together.

Art Kleiner would like to thank those at PwC and *strategy+business* who provided such strong collegiality while this book was being developed and written. These individuals include Natasha Andre, Randy Browning, Bill Cobourn, Melanie Dawe, Laura Geller, Ann-Denise Grech, Dan Gross, Gretchen Hall, Elizabeth Johnson, Paul Leinwand, Melanie Rodier, Ilona Steffen, John Sviokla, Courtney Tate, and many more; the list could be much longer. It is important to note that *The Wise Advocate* was not written or developed under the auspices of the PwC network. It does not represent the views of the network or any of its member firms. He thanks a wide range of colleagues and friends who have meant a great deal to him over the years (you know who you are). Most of all, he thanks his family: Faith, Frances, Elizabeth, Constance, and Linda. They have slowly watched this book come together, and seen its effect; and we all look forward to celebrating its publication.

Notes

1. The Choice of the Moment

1. Stephen J. Gerras, ed., *Strategic Leadership Primer*, 3rd ed. (Carlisle Barracks, Pa.: U.S. Army War College, 2010), 11.

2. James MacGregor Burns, *Transforming Leadership* (New York: Grove Press, 2004).

3. Ronald Heifetz and Marty Linsky, *Leadership on the Line* (Boston: Harvard Business Review Press, 2017); Bill George, *Authentic Leadership: Rediscovering the Secrets to Creating Lasting Value* (San Francisco: Jossey-Bass, 2003).

4. Warren Bennis and Burt Nanus, *Leaders: Strategies for Taking Charge*, 2nd ed. (New York: Collins, 2005); Warren Bennis with Patricia Ward Biederman, *The Essential Bennis* (San Francisco: Jossey-Bass, 2009); Peter G. Northouse, *Leadership*, 7th ed. (Singapore: Sage, 2016).

5. James MacGregor Burns, *Leadership* (New York: Harper & Row, 1978), 267 ff.

6. Heifetz and Linsky, *Leadership on the Line*, 14.

7. Judith Hicks Stiehm and Nicholas W. Townsend, *The U.S. Army War College: Military Education in a Democracy* (Philadelphia: Temple University Press, 2002), 6.

8. See, for example, B. M. Bass and B. J. Avioli, "Transformational Leadership: A Response to Critiques," in *Leadership Theory and Research: Perspectives and Directions*, ed. M. M. Chemers and R. Ayman (San Diego: Academic Press, 1993); Burns, *Transforming Leadership*; Robert L. Phillips and James G. Hunt, eds., *Strategic Leadership* (Westport, Conn.: Praeger, 1992); Jessica Leitch, David Lancefield, and Mark Dawson, "10 Principles of Strategic Leadership," *strategy+business*, May 18, 2016; and Jim Allen McCleskey, "Situational, Transformational, and Transactional Leadership

and Leadership Development," *Journal of Business Studies Quarterly* 5, no. 4 (June 2014): 117–130.

9. Joseph LeDoux, *The Emotional Brain: The Mysterious Underpinnings of Emotional Life* (New York: Simon & Schuster, 1996).

10. Daniel Kahneman, *Thinking, Fast and Slow* (New York: Farrar, Straus and Giroux, 2011).

11. Ann Graham, "The Thought Leader Interview: William J. O'Rourke," *strategy+business*, November 27, 2012. All quotations from O'Rourke are from this article unless otherwise noted.

12. O'Rourke, personal communication, August 2018; Brad Agle, Aaron Miller, Bill O'Rourke, *The Business Ethics Field Guide* (Provo, UT: Redtree Leadership, 2016); Frances M. Amatucci, "Alcoa Russia—Values-Driven Organizational Transformation: Case Study," retrieved April 22, 2018, https://boh.pitt.edu/sites/default/files/Alcoa%20Case%20Update%20-%206-22-17.pdf; Graham, "The Thought Leader Interview."

13. "Alcoa Might Sell One of Two Plants in Russia," *Interfax*, November 24, 2014, retrieved April 21, 2018, http://www.interfax.com/newsinf.asp?id=553397; Alcoa, "Invest in Russia," http://investinrussia.com/investors/11/pdf; VSMPO-AVISMA, "Alcoa Samara and Russia's VSMPO-AVISMA Announce Joint Venture Operational, Serve Global Aerospace Industry," September 8, 2016, http://www.vsmpo.ru/en/news/181/Arkonik_SMZ_i_VSMPOAVISMA_objavljajut_o_sozdanii_sovmestnogo_predprijatija.

14. Kevin McCallum, "How Madiba Saved the Springbok," *Cape Argus*, October 12, 2008, https://www.iol.co.za/capeargus/sport/how-madiba-saved-the-springbok-594542.

15. "How Former AT&T and GM CEO Ed Whitacre Pulled Off Two of America's Most Successful Business Turnarounds," TCU Neeley School of Business, http://www.neeley.tcu.edu/News_and_Events/Neeley_Publications/eNews/April_13/News/How_Former_AT_T_and_GM_CEO_Ed_Whitacre_Pulled_Off_Two_of_America%E2%80%99s_Most_Successful_Business_Turnarounds.aspx.

16. Sally Helgesen, "Frances Hesselbein's Merit Badge In Leadership," *strategy+business*, May 11, 2015; Thomas A. Stewart and Patricia O'Connell, "How Howard Schultz's Angel Poised Starbucks for Success," *strategy+business*, January 10, 2017; Jeffrey Rothfeder, "For Honda, Waigeya Is the Way," *strategy+business*, August 1, 2014.

17. Jeffrey M. Schwartz and Rebecca Gladding, *You Are Not Your Brain: The Four-Step Solution for Changing Bad Habits, Ending Unhealthy Thinking, and Taking Control of Your Life* (New York: Avery, 2011), 5.

18. David Rock and Jeffrey Schwartz, "The Neuroscience of Leadership," *strategy+business*, May 30, 2006; David Rock and Jeffrey Schwartz, "Why Neuroscience Matters to Executives," *strategy+business*, April 10, 2007; Jeffrey Schwartz, Pablo Gaito, and Doug Lennick, " 'That's the Way We (Used to) Do

Things Around Here,' " *strategy+business*, February 22, 2011; Jeffrey Schwartz, Josie Thomson, and Art Kleiner, "The Neuroscience of Strategic Leadership," *strategy+business*, December 5, 2016; Jeffrey Schwartz and Josie Thomson, "Conversations That Kill Your Culture," *strategy+business*, April 26, 2018.

19. Jeffrey M. Schwartz and Rebecca Gladding, *You Are Not Your Brain: The Four-Step Solution for Changing Bad Habits, Ending Unhealthy Thinking, and Taking Control of Your Life* (New York: Avery, 2011).

20. Peter M. Senge, Art Kleiner, Charlotte Roberts, Richard B. Ross, and Bryan J. Smith, *The Fifth Discipline Fieldbook: Strategies and Tools for Building a Learning Organization* (New York: Doubleday, 1994); Art Kleiner, *Who Really Matters: The Core Group Theory of Power, Privilege, and Success* (New York: Doubleday, 2003).

2. Low and High Ground

1. Reid Hoffman, "Connections with Integrity," *strategy+business*, May 29, 2012.

2. David Rowan, "Reid Hoffman: The Network Philosopher," *Wired*, April 2012.

3. Ben Casnocha, "10,000 Hours with Reid Hoffman: What I Learned," 2015, http://casnocha.com/reid-hoffman-lessons.

4. Matthew D. Lieberman, "Social Cognitive Neuroscience: A Review of Core Processes," *Annual Review of Psychology* 58 (2007): 259–289.

5. Daniel Goleman, *Emotional Intelligence: Why It Can Matter More Than IQ* (New York: Bantam, 1995), 13 ff. Chapter 2 covers the amygdala as a source of emotional hijacking; the term "amygdala hijacking" appears on page 203.

6. Charles Duhigg, *The Power of Habit: Why We Do What We Do in Life and Business* (New York: Random House, 2012).

7. Jeffrey M. Schwartz and Rebecca Gladding, *You Are Not Your Brain: The Four-Step Solution for Changing Bad Habits, Ending Unhealthy Thinking, and Taking Control of Your Life* (New York: Avery, 2011), 5.

8. Oscar Bartra, Joseph T. McGuire, and Joseph W. Kable, "The Valuation System: A Coordinate-Based Meta-analysis of BOLD fMRI Experiments Examining Neural Correlates of Subjective Value," *Neuroimage* 76 (2013): 412–427.

9. Jamil Zaki, Gilberto López, and Jason Mitchell, "Activity in Ventromedial Prefrontal Cortex Co-varies with Revealed Social Preferences: Evidence for Person-Invariant Value," *Social Cognitive and Affective Neuroscience* 9 (2014): 464–469.

10. Matthew Lieberman, "Self-Knowledge from Philosophy to Neuroscience to Psychology," Chapter 5, in *Handbook of Self Knowledge*, ed. S. Vazire and T. D. Wilson (New York: Guilford Press, 2012), 63–76.

11. Based on a biblical statement from Acts 4:35, coopted by Marxists.

12. Jack Welch and John A. Byrne, *Straight from the Gut* (New York: Warner, 2001).

13. Peter M. Senge, *The Fifth Discipline: The Art and Practice of the Learning Organization*, 2nd ed. (New York: Doubleday, 2006), 103; Daniel Kim, Michael Goodman, Jennifer Kemeny, and Charlotte Roberts, "Archetype 3: 'Shifting the Burden,' " in Peter M. Senge, Art Kleiner, Charlotte Roberts, Richard B. Ross, and Bryan J. Smith, *The Fifth Discipline Fieldbook: Strategies and Tools for Building a Learning Organization* (New York: Doubleday, 1994), 135.

14. See, for example, Jack Zenger and Joseph Folkman, "The Trickle-Down Effect of Good (and Bad) Leadership," *Harvard Business Review*, January 14, 2016; Daniel Goleman, Richard Boyatzis, and Annie McKee, *Primal Leadership: Realizing the Power of Emotional Intelligence* (Boston: Harvard Business School Press, 2002).

15. On Enron, see Bethany McLean and Peter Elkind, *The Smartest Guys in the Room: The Amazing Rise and Scandalous Fall of Enron* (New York: Portfolio, 2003). On Lehman Brothers, see Oonagh McDonald, *Lehman Brothers: A Crisis of Value* (Manchester, UK: Manchester University Press, 2015); and Vicky Ward, *The Devil's Casino: Friendship, Betrayal, and the High Stakes Games Played Inside Lehman Brothers* (New York: Wiley, 2010). On BP, see Earl Boebart and James M. Blossom, *Deepwater Horizon: A Systems Analysis of the Macondo Disaster* (Cambridge, Mass.: Harvard University Press, 2016). On Volkswagen, see Jack Ewing, *Faster, Higher, Farther: How One of the World's Largest Automakers Committed a Massive and Stunning Fraud* (New York: Norton, 2017). On Wells Fargo, see Bethany McLean, "How Wells Fargo's Cutthroat Corporate Culture Allegedly Drove Bankers to Fraud," *Vanity Fair*, summer 2017. On Equifax, see Walter Primoff, and Sidney Kess, "The Equifax Data Breach: What CPAs and Firms Need to Know Now," *CPA Journal*, December 2017.

16. Adele Diamond, "Executive Functions," *Annual Review of Psychology* 64 (2013): 135–168.

17. Walter Mischel, *The Marshmallow Test: Mastering Self-Control* (New York: Little, Brown, 2014). This is an overview of research by Mischel and others on the executive function of the brain, which is strongly linked, in our view, to the High Ground.

18. Shihui Han, Peking University, in discussion with an author of this book, October 2012.

19. Art Kleiner, "Elliott Jaques Levels with You," *strategy+business*, January 1, 2001.

20. Juan Manuel Contreras, Jessica Schirmer, Mahzarin R. Banaji, and Jason P. Mitchell, "Common Brain Regions with Distinct Patterns of

Neural Responses During Mentalizing About Groups and Individuals," *Journal of Cognitive Neuroscience* 25 (2013): 1406–1417; Jason P. Mitchell, C. Neil Macrae, and Mahzarin R. Banaji, "Dissociable Medial Prefrontal Contributions to Judgments of Similar and Dissimilar Others," *Neuron* 50 (2006): 655–663.

21. Mitchell, Macrae, and Banaji, "Dissociable Medial Prefrontal Contributions."

22. Adam Waytz, Jamil Zaki, and Jason P. Mitchell, "Response of Dorsomedial Prefrontal Cortex Predicts Altruistic Behavior," *Journal of Neuroscience* 32 (2012): 7646–7650.

23. Sylvia A. Morelli, Matthew D. Sacchet, and Jamil Zaki, "Common and Distinct Neural Correlates of Personal and Vicarious Reward: A Quantitative Meta-analysis," *NeuroImage* 112 (2015): 244–253.

24. Oscar Bartra, Joseph T. McGuire, and Joseph W. Kable, "The Valuation System: A Coordinate-Based Meta-analysis of Bold fMRI Experiments Examining Neural Correlates of Subjective Value," *Neuroimage* 76 (2013): 412–427.

25. Jeffrey Schwartz, Josie Thomson, and Art Kleiner, "The Neuroscience of Strategic Leadership," *strategy+business*, December 5, 2016.

26. Kevin N. Ochsner and James J. Gross, "The Cognitive Control of Emotion," *Trends in Cognitive Sciences* 9 (2005): 242–249.

27. Jeffrey M. Schwartz and Sharon Begley, *The Mind and the Brain: Neuroplasticity and the Power of Mental Force* (New York: Harper, 2002).

28. Kevin N. Ochsner, Silvia A. Bunge, James J. Gross, and John D. E. Gabrieli, "Rethinking Feelings: An FMRI Study of the Cognitive Regulation of Emotion," *Journal of Cognitive Neuroscience* 14 (2002): 1215–1229.

29. Jason P. Mihalik, J. Troy Blackburn, Richard M. Greenwald, Robert C. Cantu, Stephen W. Marshall, and Kevin M. Guskiewicz, "Collision Type and Player Anticipation Affect Head Impact Severity Among Youth Ice Hockey Players," *Pediatrics* 125, no. 6 (June 2010); Jaclyn B. Caccese and Thomas W. Kaminski, "Minimizing Head Acceleration in Soccer: A Review of the Literature," *Sports Medicine* 46, no. 11 (2016): 1591.

30. Ian H. Robertson and Jaap M. J. Murre, "Rehabilitation of Brain Damage: Brain Plasticity and Principles of Guided Recovery," *Psychological Bulletin* 125, no. 5 (1999): 544–575.

31. Donald Hebb, *The Organization of Behavior* (New York: Wiley, 1949); S. Löwel and W. Singer, "Selection of Intrinsic Horizontal Connections in the Visual Cortex by Correlated Neuronal Activity," *Science* 255 (1992): 209–212.

32. Jeffrey M. Schwartz, *Brain Lock: A Four-Step Self Treatment Method to Change Your Brain Chemistry* (New York: HarperCollins, 1996).

33. Schwartz and Begley, *The Mind and the Brain*; Jeffrey M. Schwartz, Henry P. Stapp, and Mario Beauregard, "Quantum Physics in Neuroscience

and Psychology: A Neurophysical Model of the Mind-Brain Interaction," *Philosophical Transactions of the Royal Society, B: Biological Sciences* 360 (2005): 1309–1327.

34. Schwartz, Stapp, and Beauregard, "Quantum Physics in Neuroscience and Psychology"; Hebb, *The Organization of Behavior*.

3. Finding Your Inner Voice

1. Adam Smith, *An Inquiry Into the Nature and Causes of the Wealth of Nations* (Edinburgh: William Strahan, 1776).

2. Adam Smith, *The Theory of Moral Sentiments*, 6th ed., ed. D. D. Raphael and A. L. Macfie (Indianapolis: Liberty Fund, 1982), 292. [1st ed. pub. 1759; 6th ed. pub. 1790]

3. Smith, *The Theory of Moral Sentiments*, 134.

4. Ryan Patrick Hanley, "Living a Life," in *Adam Smith: His Life, Thought and Legacy*, ed. Ryan Patrick Hanley (Princeton, N.J.: Princeton University Press, 2016).

5. Jeffrey M. Schwartz and Rebecca Gladding, *You Are Not Your Brain: The Four-Step Solution for Changing Bad Habits, Ending Unhealthy Thinking, and Taking Control of Your Life* (New York: Avery, 2011), 7.

6. Carlo Collodi, *Le avventure di Pinocchio* (1883), trans. M. A. Murray (London: T. Fisher Unwin, 1892). The character, called only the "talking cricket," was named Jiminy Cricket for the 1940 animated film *Pinocchio*.

7. 1 Kings 19:12.

8. Hanley Ryan Patrick, *Adam Smith and the Character of Virtue* (Cambridge: Cambridge University Press, 2009), 140–141.

9. Chris D. Frith and Uta Frith, "Mechanisms of Social Cognition," *Annual Review of Psychology* 63 (2012): 287–313.

10. Chris D. Frith and Uta Frith, "The Neural Bases of Mentalizing," *Neuron* 50 (2006): 531–534; Brian T. Denny, Hedy Kober, Tor D. Wager, and Kevin N. Ochsner, "A Meta-analysis of Functional Neuroimaging Studies of Self- and Other Judgments Reveals a Spatial Gradient for Mentalizing in Medial Prefrontal Cortex," *Journal of Cognitive Neuroscience* 24:8 (August 2012): 1742-1752.

11. Juan Manuel Contrera, Jessica Schirmer, Mahzarin R. Banaji, and Jason P. Mitchell, "Common Brain Regions with Distinct Patterns of Neural Responses During Mentalizing About Groups and Individuals," *Journal of Cognitive Neuroscience* 25 (2013), 1406–1417.

12. Ram Charan, "Why the Aetna-CVS Deal Is a Lesson for Leaders," *strategy+business*, December 7, 2017.

13. Jon Katzenbach, Gretchen Anderson, and Art Kleiner, "Mark Bertolini's Preventive Disruption," *strategy+business*, April 13, 2015. Subsequent quotes from Mark Bertolini are all from this article unless otherwise noted.

14. See, for example, Kimberly Leonard, "Aetna CEO on Obamacare: 'We Can Fix It,'" *Washington Examiner*, April 29, 2018.

15. Katzenbach, Anderson, and Kleiner, "Mark Bertolini's Preventive Disruption."

16. Peter M. Senge, *The Fifth Discipline: The Art and Practice of the Learning Organization*, 2nd ed. (New York: Doubleday, 2006), 172 ff; Art Kleiner, *The Age of Heretics: A History of the Radical Thinkers Who Reinvented Corporate Management* (San Francisco: Jossey-Bass, 2010), 215 ff; Chris Argyris, *Overcoming Organizational Defenses* (Needham Heights, Mass.: Allyn & Bacon, 1990).

17. Jamil Zaki, Jochen Weber, Niall Bolger, and Kevin Ochsner, "The Neural Bases of Empathic Accuracy," *Proceedings of the National Academy of Sciences* 106 (2009): 11382–11387.

18. Sylvia A. Morelli, Matthew D. Sacchet, and Jamil Zaki, "Common and Distinct Neural Correlates of Personal and Vicarious Reward: A Quantitative Meta-analysis," *NeuroImage* 112 (2015): 244–253.

19. Keely A. Muscatell, Sylvia A. Morelli, Emily B. Falk, Baldwin M. Way, Jennifer H. Pfeifer, Adam D. Galinsky, Matthew D. Lieberman, Mirella Dapretto, and Naomi I. Eisenberger, "Social Status Modulates Neural Activity in the Mentalizing Network," *NeuroImage* 60 (2012): 1771–1777; David M. Amadio and Chris D. Frith, "Meeting of Minds: The Medial Frontal Cortex and Social Cognition," *Nature Reviews Neuroscience* 7 (2006): 268–277; Frith and Frith, "Mechanisms of Social Cognition."

20. Emily B. Falk, Sylvia A. Morelli, B. Locke Welborn, Karl Dambacher, and Matthew D. Lieberman, "Creating Buzz: The Neural Correlates of Effective Message Propagation," *Psychological Sciences* 24 (2013):1234–1242; Matthew Brook O'Donnell, Emily B. Falk, and Matthew D. Lieberman, "Social In, Social Out: How the Brain Responds to Social Language with More Social Language," *Communication Monographs* 82 (2015): 31–63.

21. Muscatell et al., "Social Status Modulates Neural Activity."

22. Muscatell et al., "Social Status Modulates Neural Activity"; Amadio and Frith, "Meeting of Minds"; Frith and Frith, "Mechanisms of Social Cognition."

23. Robert Fuller, *Somebodies and Nobodies: Overcoming the Abuse of Rank* (Gabriola Island, B.C.: New Society, 2003).

24. Muscatell et al., "Social Status Modulates Neural Activity."

25. Robert K. Greenleaf, *The Servant as Leader*, rev. ed. (Westfield, Ind.: Greenleaf Center for Servant Leadership, 2008).

26. Naomi I. Eisenberger, Tristen K. Inagaki, Keely A. Muscatell, Kate E. Byrne Haltom, and Mark R. Leary, "The Neural Sociometer: Brain Mechanisms Underlying State Self-Esteem," *Journal of Cognitive Neuroscience* 23 (2011): 3448–3455.

27. Susan T. Fiske and Eric Dépret, "Control, Interdependence and Power: Understanding Social Cognition in Its Social Context," *European Review of Social Psychology* 7 (2011): 31–61.

28. See, for examples, Jerry Useem, "Power Causes Brain Damage," *Atlantic Monthly*, July/August 2017; and David Owen and Jonathan Davidson, "Hubris Syndrome: An Acquired Personality Disorder? A Study of US Presidents and UK Prime Ministers Over the Last 100 Years," *Brain* 132, no. 5 (May 2009): 1396–1406.

29. Christina Novicki, "Don't Change That Channel . . . Change the Rules!," *Fast Company*, December 31, 1996.

30. Paul Leinwand and Cesare Mainardi, *Strategy That Works: How Winning Companies Close the Strategy-to-Execution Gap* (Boston: Harvard Business Review Press, 2016), 24.

31. Art Kleiner, "Thought Leaders: Thomas Malone on Building Smarter Teams," *strategy+business*, May 12, 2014; Enrique G. Fernández-Abascal, Rosario Cabello, Pablo Fernández-Berrocal, and Simon Baron-Cohen, "Test-Retest Reliability of the 'Reading the Mind in the Eyes' Test: A One-Year Follow-up Study," *Molecular Autism* 4 (2013): 33; Simon Baron-Cohen, *Mindblindness: An Essay on Autism and Theory of Mind* (Boston: MIT Press, 1995).

32. Charles Duhigg, "What Google Learned from Its Quest to Build the Perfect Team," *New York Times*, February 26, 2016.

33. Kleiner, "Thought Leaders: Thomas Malone on Building Smarter Teams."

34. J. R. R. Tolkien, *The Two Towers: Being the Second Part of the Lord of the Rings* (New York: Houghton-Mifflin, 1954), 205.

35. Adele Diamond, "Executive Functions," *Annual Review of Psychology* 64 (2013): 136.

36. Diamond, "Executive Functions."

37. Diamond, "Executive Functions."

38. Carol S. Dweck, *Mindset: The New Psychology of Success* (New York: Ballantine Books, 2007); Carol S. Dweck and Ellen L. Leggett, "A Social-Cognitive Approach to Motivation and Personality," *Psychological Review* 95 (1988): 256–273.

39. James J. Gross, "Emotion Regulation: Current Status and Future Prospects," *Psychological Inquiry* 26:1 (2015), 1-26; James J. Gross, "Emotion Regulation: Affective, Cognitive, and Social Consequences," *Psychophysiology* 39 (2002): 281–291.

40. Antoine Lutz, Heleen A. Slagter, John D. Dunne, and Richard J. Davidson, "Attention Regulation and Monitoring in Meditation," *Trends in Cognitive Sciences* 12 (2008): 163–169.

41. Norman A. S. Farb, Zindel V. Segal, Helen Mayberg, Jim Bean, Deborah McKeon, Zainab Fatima, and Adam K. Anderson, "Attending to the Present: Mindfulness Meditation Reveals Distinct Neural Modes of Self-Reference," *Social Cognitive and Affective Neuroscience* 2 (2007): 313–322.

42. Antoine Lutz, Lawrence L. Greischar, Nancy B. Rawlings, Matthieu Ricard, and Richard J. Davidson, "Long-Term Meditators Self-Induce High-Amplitude Gamma Synchrony During Mental Practice," *Proceedings of the National Academy of Sciences* 101 (2004): 16369–16373.

43. Lutz et al., "Long-Term Meditators"; John Kounios and Mark Beeman, "The Cognitive Neuroscience of Insight," *Annual Review of Psychology* 65 (2014): 71–93.

44. Wendy Hasenkamp and Lawrence W. Barsalou, "Effects of Meditation Experience on Functional Connectivity of Distributed Brain Networks," *Frontiers of Human Neuroscience* 6:38 (2012): 1-14.

45. John H. Flavell, "Metacognition and Cognitive Monitoring: A New Area of Cognitive-Developmental Inquiry," *American Psychologist* 34 (1979): 906–911.

46. Wendy Hasenkamp, Christine D. Wilson-Mendenhall, Erica Duncan, and Lawrence W. Barsalou, "Mind Wandering and Attention During Focused Meditation: A Fine-Grained Temporal Analysis of Fluctuating Cognitive States," *NeuroImage* 59 (2012): 750–760.

47. William W. Seeley, Vinod Menon, Alan F. Schatzberg, Jennifer Keller, Gary H. Glover, Heather Kenna, Allan L. Reiss, and Michael D. Greicius, "Dissociable Intrinsic Connectivity Networks for Salience Processing and Executive Control," *Journal of Neuroscience* 27 (2007): 2349–2356.

48. Hasenkamp and Barsalou, "Effects of Meditation Experience," 28.

49. Farb et al., "Attending to the Present."

50. Hans Helmut Kornhuber and Lüder Deecke, *The Will and Its Brain: An Appraisal of Reasoned Free Will* (Lanham, Md.: University Press of America, 2012).

51. Benjamin Libet, "Do We Have Free Will?," *Journal of Consciousness Studies* 6 (1999): 47–57.

52. Jeffrey Schwartz and Sharon Begley, *The Mind and the Brain: Neuroplasticity and the Power of Mental Force* (New York: Harper Perennial, 2003), chap. 9.

53. Diamond, "Executive Functions."

54. Libet, "Do We Have Free Will?"

55. Schwartz and Begley, *The Mind and the Brain*, 306.

56. Diamond, "Executive Functions."

57. Micah Allen, Martin Dietz, Karina S. Blair, Martijn van Beek, Geraint Rees, Peter Vestergaard-Poulsen, Antoine Lutz, and Andreas Roepstorff, "Cognitive-Affective Neural Plasticity Following Active-Controlled Mindfulness Intervention," *Journal of Neuroscience* 32 (2012): 15601–15610.

58. Norman Farb and Wolf E. Mehling, "Editorial: Interoception, Contemplative Practice, and Health," *Frontiers in Psychology*, December 1, 2016, doi:10.3389/fpsyg.2016.01898.

59. Allen et al., "Cognitive-Affective Neural Plasticity."

60. Norman A. S. Farb, Zindel V. Segal, and Adam K. Anderson, "Mindfulness Meditation Training Alters Cortical Representations of Interoceptive Attention," *Social Cognitive and Affective Neuroscience* 8 (2013):15–26; Farb and Mehling, "Editorial."

61. Shihui Han, Lihua Mao, Xiaosi Gu, Ying Zhu, Jianqiao Ge, and Yina Ma, "Neural Consequences of Religious Belief on Self-Referential Processing," *Social Neuroscience* 3 (2008): 1–15.

62. The words were derived from Yufan Liu, *Modern Lexicon of Chinese Frequently Used Word Frequency*, (Beijing, China: Space Navigation Press, 1990).

63. Han et al, "Neural Consequences."

4. Relabeling Your Messages

1. Chris Argyris, "Teaching Smart People How to Learn," *Harvard Business Review* 69, no. 3 (May 1991): 99–109.

2. William R. Noonan, *Discussing the Undiscussable: A Guide to Overcoming Defensive Routines in the Workplace* (San Francisco: Jossey-Bass, 2007).

3. Carmen Reinhart and Kenneth Rogoff, *This Time Is Different: Eight Centuries of Financial Folly* (Princeton, N.J.: Princeton University Press, 2009), 1.

4. Carol S. Dweck and Elaine S. Elliott-Moskwa, "Self-Theories: The Roots of Defensiveness," in *Social Psychological Foundations of Clinical Psychology*, ed. James E. Maddux and June Price Tangney (New York: Guilford Press, 2010).

5. Albert Bandura, *Self-Efficacy: The Exercise of Control* (New York: Worth, 1997).

6. Daniel Kahnemann, *Thinking Fast and Slow* (New York: Farrar, Strauss and Giroux, 2011), 282ff.

7. Richard B. Ross, "The Ladder of Inference," in Peter M. Senge, Art Kleiner, Charlotte Roberts, Richard B. Ross, and Bryan J. Smith, *The Fifth*

Discipline Fieldbook: Strategies and Tools for Building a Learning Organization (New York: Doubleday, 1994), 242.

8. David Kantor, *Reading the Room: Group Dynamics for Coaches and Leaders* (San Francisco: Jossey-Bass, 2012), 40–41.

9. Ann Graham, "The Company That Anticipated History," *strategy+business*, November 30, 2006.

10. Ian McRae, "Electricity for All," in Peter M. Senge, Charlotte Roberts, Art Kleiner, Richard B. Ross, George Roth, and Bryan Smith, *The Dance of Change: The Challenges to Sustaining Momentum in a Learning Organization* (New York: Crown Business, 1999), 541.

11. McRae, "Electricity for All," 541.

12. Jon Katzenbach, Gretchen Anderson, and Art Kleiner, "Mark Bertolini's Preventive Disruption," *strategy+business*, April 13, 2015.

13. Art Kleiner and George Roth, *Oil Change: Perspectives on Corporate Transformation* (Oxford: Oxford University Press, 2000), 109.

5. Reframing Your Situation

1. Art Kleiner and Josie Thomson, "Sovereign Wealth Funds Can Be Management Leaders" (interview with Damian Frawley and Marcus Simpson of QIC), *strategy+business*, January 13, 2016. Subsequent quotations are from this article unless otherwise noted.

2. Carmen Reinhart and Kenneth Rogoff, *This Time Is Different: Eight Centuries of Financial Folly*, (Princeton, N.J.: Princeton University Press, 2009).

3. Adam Gopnik, "Bumping Into Mr. Ravioli: A Theory of Busyness, and Its Hero," *New Yorker*, September 30, 2002, 80.

4. Jesse Sostrin " 'Yes' vs. 'Yes, If . . .': Using Your Distinctive Contribution to Manage Priorities," *strategy+business*, May 2, 2016.

5. Jeffrey Schwartz and Josie Thomson, "Conversations That Kill Your Culture," *strategy+business*, April 26, 2018.

6. Jeffrey Schwartz, Pablo Gaito, and Doug Lennick, "That's The Way We (Used to) Do Things Around Here," *strategy+business*, February 22, 2011.

7. Conversation with two of the authors, July 2016.

8. Daniel J. Tomasulo and Daniel Lerner, "The Happiness Advantage: An Interview with Shawn Achor," PsychCentral, 2012, accessed April 23, 2018, https://psychcentral.com/blog/the-happiness-advantage-an-interview-with-shawn-achor/.

9. Donald D. Price, Damien G. Finniss, and Fabrizio Benedetti, "A Comprehensive Review of the Placebo Effect: Recent Advances and Current Thought," *Annual Review of Psychology* 59 (2008): 565-90.

10. Tetsuo Koyama, John G. McHaffie, Paul J. Laurienti and Robert Coghill, "The Subjective Experience of Pain: Where Expectations Become Reality," *Proceedings of the National Academy of Sciences* 102:36 (September 2005): 12950-12955; Robert C. Coghill, "Individual Differences in the Subjective Experience of Pain: New Insights Into Mechanisms and Models," *Headache* 50 (2010): 1531–1535.

11. Conversation with one of the authors, January 2014. All quotations from Bonnano are from this conversation unless otherwise noted.

6. Refocusing Your Attention

1. Conversation with one of the authors.

2. Quoted in Alan Deutschman, *Change or Die: The Three Keys to Change at Work and in Life* (New York: HarperCollins, 2008), 4.

3. Conversation with one of the authors, July, 2015.

4. Marshall Goldsmith and Mark Reiter, *Triggers: Creating Behavior That Lasts, Becoming the Person You Want to Be* (New York: Crown Business, 2015).

5. B. Misra and E. C. G. Sudarshan, "The Zeno's Paradox in Quantum Theory," *Journal of Mathematical Physics* 18 (1977): 756–763.

6. Jeffrey M. Schwartz, Henry P. Stapp, and Mario Beauregard, "Quantum Theory in Neuroscience and Psychology: A Neurophysical Model of Mind/ Brain Interaction," *Philosophical Transactions of the Royal Society B* 360 (2005): 1309–1327.

7. Yi-Yuan Tang, Britta K. Holzel, and Michael I. Posner, "The Neuroscience of Mindfulness Meditation," *Nature Reviews Neuroscience* 16 (2015): 213–225.

8. Adele Diamond, "Executive Functions," *Annual Review of Psychology* 64 (2013): 135–168.

9. Conversation with one of the authors. All quotes from this individual come from that same conversation.

10. The Hunker'd Down Blues Band website is at https://hunkerddown bluesband.bandcamp.com/.

11. Jeffrey Schwartz, Pablo Gaito, and Doug Lennick, "That's The Way We (Used to) Do Things Around Here," *strategy+business*, February 22, 2011.

12. Peter M. Senge, *The Fifth Discipline: The Art and Practice of the Learning Organization*, 2nd ed. (New York: Doubleday, 2006), 139.

13. Sylvia A. Morelli, Matthew D. Sacchet, and Jamil Zaki, "Common and Distinct Neural Correlates of Personal and Vicarious Reward: A Quantitative Meta-analysis," *NeuroImage* 112 (2015): 244–253; Adam Waytz, Jamil Zaki, and Jason P. Mitchell, "Response of Dorsomedial Prefrontal Cortex Predicts Altruistic Behavior," *Journal of Neuroscience* 32 (2012): 7646–7650; Jamil Zaki, "Empathy: A Motivated Account," *Psychological Bulletin* 140 (2014): 1608–1647.

14. Conversations with the authors, all in 2016.

15. Alan Carr, *The Easy Way to Stop Smoking* (London: Arcturus, 1985); "Alan Carr, Saviour of Smokers, Died on November 29th, Aged 72," *Economist*, December 7, 2006.

16. Conversations with the authors, July 2016.

17. David Welch, "Ed Whitacre's Battle to Save GM from Itself," *Bloomberg Business Week*, April 29, 2010.

18. Art Kleiner, "Jack Stack's Story Is an Open Book," *strategy+business*, July 1, 2001.

7. Revaluing Your Leadership

1. The phrase, apparently of Quaker origin, represents in our view a longstanding religious and ethical tradition aligned with the Wise Advocate: voicing broader moral perspectives to those in positions of authority in a way that they can hear. See John Green, "The Origin of the Phrase 'Speaking Truth to Power,'" Classroom, https://classroom.synonym.com/origin-phrase -speaking-truth-power-11676.html.

2. Art Kleiner and Josie Thomson, "Overcoming Challenges During a Major Transformation," *strategy+business*, February 21, 2018. Subsequent quotes from Michael Pennisi are all from this article unless otherwise noted.

3. Joel Kurtzman, "Is Your Company Off Course?," *Fortune*, February 17, 1997; Philip J. Carroll, *Rebounding, Rebuilding, Renewing at Shell Oil* (Cambridge, Mass.: Pegasus Communications, 1999); Phil Carroll, "The Executive Leader's Perspective," in Peter M. Senge, Charlotte Roberts, Art Kleiner, Richard B. Ross, George Roth, and Bryan Smith, *The Dance of Change: The Challenges to Sustaining Momentum in Learning Organizations* (New York: Doubleday, 1999), 203.

4 Linda Pierce, "A Personal Checklist for Leaders Facing 'Messy' Experiences," in Senge et al., *The Dance of Change*, 211.

5. Art Kleiner and Josie Thomson, "Sovereign Wealth Funds Can Be Management Leaders," *strategy+business*, January 13, 2016.

6. Art Kleiner, *Who Really Matters: The Core Group Theory of Power, Privilege, and Success* (New York: Doubleday, 2003).

7. Conversations with an author, September 2013-January 2014. All subsequent quotes from Noel Lord are from the same group of conversations.

8. Conversations with an author, April 2017.

9. Art Kleiner, *The Age of Heretics: A History of the Radical Thinkers Who Reinvented Corporate Management*, 2nd ed. (San Francisco: Jossey-Bass, 2010).

200   *7. Revaluing Your Leadership*

10. Conversations with the authors, July 2016.

11. Peter Harry Brown and Pat H. Broeske, *Howard Hughes: The Untold Story* (Boston: Da Capo Press, 1996).

12. Donald L. Barlett and James B. Steele, *Howard Hughes: His Life and Madness* (New York: Norton, 2004), chap. 6.

13. Barlett and Steele, *Howard Hughes*, chap. 4 and 5.

14. George J. Marrett, *Howard Hughes: Aviator* (Annapolis, Md.: Naval Institute Press, 2004), 62; Barlett and Steele, *Howard Hughes*, chap. 4 and 5.

15. Naomi I. Eisenberger, Tristen K. Inagaki, Keely A. Muscatell, Kate E. Byrne Haltom, and Mark R. Leary, "The Neural Sociometer: Brain Mechanisms Underlying State Self-Esteem," *Journal of Cognitive Neuroscience* 23 (2011), 3448–3455.

16. Conversation with one of the authors.

17. Jeffrey M. Schwartz, *A Return to Innocence* (New York: ReganBooks, 1998); later revised as *Dear Patrick: Life Is Hard—Here's Some Good Advice* (New York: ReganBooks, 2010).

18. Kleiner, *The Age of Heretics*.

Glossary

Applied mindfulness: a state of mental activity which combines meta-cognition, or thinking about your thinking, with meta-attention, or paying attention to how you pay attention. Applied mindfulness is common to both High Ground and Higher Ground mental activity. For the High Ground, it involves consulting the Wise Advocate and mentalizing about yourself. This is the form of mindfulness that most initial practitioners follow. For the Higher Ground, it is a gateway to deep mindfulness and meta-awareness. (Chapter 3.)

Attention density: the repeated activity of making choices (whether wise or foolish) about where and how you focus your attention. When these choices are sustained in consistent ways, over time, you rewire the circuits of neural activity in your brain (for better or for worse). Also see Refocusing. (Chapters 2 and 6.)

Brain: the biological organ, located in the skull, where neurons and other cells process and transmit sensations, feelings, thoughts, and emotions. The brain is affected by the mind's choices and decisions. The more capable you become at recognizing and managing your brain's reactions, the easier it becomes to focus your attention effectively. (Chapter 1.)

Cognitive distortions: a synonym for deceptive brain messages; at the level of organizational conversation and culture, these are cognitive organizational distortions. (Chapters 1 and 4.)

Cognitive flexibility: the ability to shift perspectives and adapt to new challenges and opportunities as they arise. This is a key attribute of executive function, and thus of the High Ground and strategic leadership. (Chapter 3.)

Core group: the people that others in an organization think about when they make decisions, and on whose behalf they set priorities. The core group tends to include people in authority; it often also includes others, lower in the hierarchy, who have legitimacy and respect from the rest of the enterprise. When you revalue your leadership or move toward Higher Ground, it enhances your ability to play the role of Wise Advocate for or with the core group. (Chapter 7.)

Deceptive brain messages: thoughts, feelings and sensations that arise into consciousness, generated by habitual, near-automatic brain processes. If you're typical, you receive a steady stream of these all through your waking hours. They are incomplete depictions of reality that seem as though their complete, and if they are not examined, they can lead people to make poor decisions. Also see Relabeling. (Chapters 1 and 4.)

Deceptive organizational messages: the collective equivalent of deceptive brain messages. They are embedded in an organization's culture, and manifest in everyday conversation. Chapter 4 contains a list of common deceptive organizational messages. (Chapter 4.)

Deep mindfulness: a profound, practiced form of mindfulness in which you are aware of how you are consulting the Wise Advocate. This is associated with the Higher Ground, and one of its characteristics is a high level of meta-awareness. (Chapter 3.)

Deliberative Self-Referencing Center: a function associated with the High Ground. Activated by mentalizing, it involves considerations of what people are thinking and what they might do next. In the physical brain, it is located in the dorsal (upper) medial prefrontal cortex (dmPFC), which sits just above the part of the brain where the Reactive Self-Referencing Center is located. (Chapter 2.)

Emotional reasoning: a common kind of deceptive (brain or organizational) message in which emotional responses are mistaken for rational thought. For example, if you feel afraid about a potential opportunity, this might lead to excessive risk aversion. (Chapter 4.)

Excessive risk aversion: Also known as "catastrophizing;" a tendency to ruminate on deceptive brain messages about worst-case scenarios. (Chapter 4.)

Executive Center: a function associated with the lateral prefrontal cortex and with executive function. It is active in the High Ground. (Chapter 2.)

Executive function: a generally accepted term for the mental activity associated with reasoning, problem-solving, and goal-directed planning. This function is a key element of strategic leadership. Three core attributes of executive function are inhibitory control (mastering emotions and impulses, associated with free won't), cognitive flexibility (being able to change direction readily), and a high-capacity working memory (the ability to keep relatively complex ideas in mind). This function is associated with the Executive Center. (Chapter 3.)

Free won't: the ability to recognize brain-based impulses and cravings without giving in to them or acting on them. Anyone can strengthen this form of inhibitory control, for example through applied mindfulness and (more specifically) through relabeling, reframing and refocusing. (Chapters 3, 4 and 6.)

Habit: a recurring pattern of behavior or mental activity (often automatic or near-automatic), that tends to occur at least partly below the level of conscious awareness. This behavior is associated with deeply entrenched neural pathways. As the habitual activity is repeated, the pathways associated with it become even more deeply entrenched, and thus the habit becomes easier to practice and harder to resist. The free won't function is often valuable as a way to resist maladaptive habits. (Chapter 2.)

Habit Center: a function active in the Low Ground. Associated with the basal ganglia and automatic behaviors and responses, the Habit Center manifested itself early in animal evolution. It is sometimes called the "lizard brain." (Chapter 2.)

Hebb's Law: A core insight expressed by Canadian scientist Donald Hebb: "Neurons that fire together wire together." Parts of the brain that are continually activated together (such as the circuits of the Low or High Ground) will physically associate with one another in the future. This describes the basic dynamic of neuroplasticity. Also see the Quantum Zeno Effect. (Chapter 2.)

High Ground: a pattern of mental activity, and the corresponding brain circuits, that is a source of strategic leadership. This pattern is invoked by mentalizing, applied mindfulness, and executive function; it is associated with the Executive Center, Deliberative Self-Referencing Center, and Warning Center. (Chapter 2.)

High road: one of two pathways identified by Joseph LeDoux that influence human behavior. The high road, in which impulses move from the thalamus to the neocortex and then to the amygdala, involves relatively slow and thoughtful responses to external events. Our Low Ground, High Ground, and Higher Ground are all associated with LeDoux's high road. (Chapter 1.)

Higher Ground: a pattern of mental activity that is a source of transformational leadership. This pattern is invoked by deep mindfulness and by repeated attention to the Wise Advocate. It links the Habit and Executive Centers, so that the executive function becomes habitual. (Chapter 3.)

Immature leadership: activity by a person of influence that reflects the individual leader's emotional responses, and typically does not help move an organization forward. This is associated with the Lower Ground. The leader expects legitimacy and receives it so long as people perceive it to be in their interest to grant it. (Chapter 3.)

Impartial spectator: A form of mental activity proposed by Adam Smith to ameliorate the ethical problems of market economics. This involves cultivating an inner voice representing the point of view of a well-informed and genuinely mindful outsider. Cultivating the impartial spectator is complementary to calling on the Wise Advocate. (Chapter 3.)

Impulse loyalty: reluctance to abandon familiar, comfortable cravings and thoughts—a habit that keeps people from moving toward the High Ground and becoming strategic leaders. (Chapter 6.)

Ladder of Inference: a conceptual device, originally developed by S. I. Hayakawa, that makes explicit the mental activity involved in assumptions and deceptive messages. It is useful in relabeling. Also see Organizational Learning. (Chapter 4.)

Legitimacy: The quality of being perceived to be worthy of respect and allegiance. The core group members of an organization are those viewed as having legitimacy. Playing the role of a Wise Advocate, if you do it effectively, makes you more likely to acquire this type of legitimacy. (Chapter 7.)

Low Ground: a pattern of mental activity, and its corresponding brain circuits, linked with transactional leadership. This pattern is invoked by thinking about what people want and by decisions on behalf of expedience; it is associated with the Habit Center, Warning Center, and Reactive Self-Referencing Center. (Chapter 2.)

Low road: one of two pathways identified by Joseph LeDoux that influence human behavior. The low road, in which impulses move directly from the thalamus to the amygdala, involves relatively rapid and emotionally charged responses to external events. Our Lower Ground (impulses and emotion) pattern of activity is associated with LeDoux's low road. (Chapter 1.)

Lower Ground: a pattern of mental activity, and the corresponding brain circuits, linked with immature leadership. In this pattern, decisions are driven by instinct and emotion. (Chapter 3.)

Mentalizer's Paradox: The observed phenomenon that as leaders rise in a hierarchy and gain social status , they are less likely to mentalize, which makes them less likely to cultivate the High Ground and become strategic leaders. (Chapter 3.)

Mentalizing: reflecting on what other people are thinking and what they are likely to do next. Mentalizing invokes the High Ground. Also see Deliberative Self-Referencing Center. (Chapter 3.)

Meta-attention: paying attention to the way you are paying attention. This is a component of applied mindfulness. It is characteristic of the High Ground and a gateway to the Higher Ground. (Chapter 3.)

Meta-awareness: becoming aware of your own awareness. The inner question, "Am I consulting my Wise Advocate?" is a hallmark of being in this state. It generally requires a great deal of practice to consistently reach meta-awareness; it is characteristic of the Higher Ground. (Chapter 3.)

Meta cognition: thinking about your own thinking. This is a component of applied mindfulness. Like meta-attention, it is characteristic of the High Ground and a gateway to the Higher Ground. (Chapter 3.)

Mind: an ongoing pattern of mental activity associated with the self, and especially with the conscious choices and decisions made by any human being. The mind is active; it's the source of choices and decisions you make about where and how to focus your attention. (Chapter 1.)

Neuroplasticity: the ability of neurons to forge new connections in response to environmental change. Parts of the brain that are continually activated together will physically associate with one another in the future. Also see Hebb's Law. (Chapter 2.)

Organizational learning: a body of theory and practice that can help members of a group move to the High and Higher Ground and become more strategic and transformational as leaders. The steps of relabeling, reframing, refocusing and revaluing are congruent with these practices. The Ladder of Inference is one of the best-known practices in organizational learning. (Chapters 4 and 7.)

Quantum Zeno Effect: A principle of physics that explains the value and power of focused attention. When any system is observed in a rapid, repetitive fashion, it affects the probabilities of change. The act of focusing close attention on your mental experience stabilizes the brain state arising in association with that experience , and brings Hebb's Law into play. (Chapter 6.)

Reactive Self-Referencing Center: a function associated with the Low Ground. Activated by expedient problem-solving, it involves consideration of what you and other people want and how to get it. In the physical brain, it is located in the ventral (lower) medial prefrontal cortex (vmPFC), which sits just below the part of the brain where the Deliberative Self-Referencing Center is located. (Chapter 2.)

Refocus: to place your attention, again and again, on the behaviors, practices, and mental activities you have chosen to emphasize. Practiced regularly, this invokes self-directed neuroplasticity and can help influence an organization for the better. This is the third step of strategic leadership. Also see Quantum Zeno Effect. (Chapter 6.)

Reframe: to explicitly replace deceptive brain or organizational messages with more constructive narratives you have chosen. This can help set a positive direction for you and your organization. It is the second step of strategic leadership. (Chapter 5.)

Relabel: to raise awareness of the nature of deceptive messages that affect you and your organization. In relabeling, you articulate the nature of these messages as coming from the brain's physical activity or the organization's habitual culture – and thus not necessarily representing reality accurately. This is the first step of strategic leadership. (Chapter 4.)

Revalue: to raise your awareness of the value of your own leadership in light of your increasing access to your inner Wise Advocate, and thus to develop your ability to play a Wise Advocate role in the organization around you. This is the fourth step of strategic leadership. Also see Core group. (Chapter 7.)

Salience: the quality of attracting attention by being striking and conspicuous. Attention tends to flow to whatever is most salient, which often leads people to the Low Ground. An exception is applied mindfulness, which can make elements of the High Ground or Higher Ground more salient. (Chapter 3.)

Self-directed neuroplasticity: the deliberate focusing of attention in a way that reorients the physical circuits of your brain. Also see Neuroplasticity. (Chapter 2.)

Self-Referencing Center: a function associated with the medial prefrontal cortex, involved in many aspects of personality and identity. (Chapter 2.)

Strategic leadership: activity by a person of influence that helps address problems that are too complex for an organization to manage easily. This is associated with the High Ground. The leader gains legitimacy by helping the organization transcend its limits and move toward more profound and powerful levels of capability. (Chapters 1 and 7.)

Subjective valuation: Mental activity concerned with what is valuable and relevant. This is closely associated with the Low Ground. Also see Reactive Self-Referencing Center. (Chapter 2.)

Transactional leadership: activity by a person of influence in which problems are solved with a high level of expediency. This is associated with the Low Ground. The leader retains legitimacy so long as peoples' wants and needs are gratified. (Chapter 2.)

Transformational leadership: activity by a person of influence that helps an organization make the most of its potential. This is associated with the Higher Ground. The leader assumes legitimacy by playing the role of Wise Advocate for the organization, and helping it become what it is called to be. (Chapters 3 and 7.)

True self: your core of personal belief and identity, as evoked by your Wise Advocate. Your true self continually becomes clearer, more relevant, and more fully developed throughout your life as a leader. It represents your sincere striving to embody the values and achieve the goals you fundamentally believe in. Strategic and transformational leaders have inevitably spent time getting in touch with their true self. (Chapter 1.)

Warning Center: a function associated with three parts of the brain: the amygdala, the insula, and the orbital frontal cortex. This center generates feelings of fear, gut-level responses, and the sense that something is worth pursuing or avoiding. It is connected to both the High and Low Ground. (Chapter 2.)

Wise Advocate: an aspect of your attentive mind that can see the
bigger picture, including your inherent worth, capabilities and
accomplishments as an individual; and your organization's most
promising potential. The Wise Advocate wants what is best for the
whole system in which you are involved, including what is best
for you according to the values and interests of your true self.
It approaches your authentic emotions and needs from a loving,
caring, nurturing and forthright perspective. As a strategic or transformational leader, you can play this role for the larger organization.
(Chapters 1–7.)

Selected Bibliography

Buber, Martin. *I and Thou*. Trans. Ronald Gregor Smith. New York: Charles Scribner's Sons, 1937.

Burns, James MacGregor. *Leadership*. New York: Harper & Row, 1978.

——. *Transforming Leadership*. New York: Grove Press, 2004.

Deutschman, Alan. *Change or Die: The Three Keys to Change at Work and in Life*. New York: HarperCollins, 2008.

Duckworth, Angela. *Grit: The Power of Passion and Perseverance*. New York: Scribner, 2016.

George, Bill. *Authentic Leadership: Rediscovering the Secrets to Creating Lasting Value*. San Francisco: Jossey-Bass, 2003.

Goleman, Daniel, *Emotional Intelligence: Why It Can Matter More Than IQ*. New York: Bantam, 1995.

Grass, Günter. *Crabwalk*. Göttingen: Steidl, 2002.

Heifetz, Ronald A. *Leadership Without Easy Answers*. Cambridge, Mass.: Harvard University Press, 1998.

Heifetz, Ronald A., and Marty Linsky. *Leadership on the Line*. Boston: Harvard Business Review Press, 2017.

Kahneman, Daniel. *Thinking, Fast and Slow*. New York: Farrar, Straus and Giroux, 2011.

Kleiner, Art. *The Age of Heretics: A History of the Radical Thinkers Who Reinvented Corporate Management*. San Francisco: Jossey-Bass, 2010.

Kleiner, Art, and George Roth. *Oil Change: Perspectives on Corporate Transformation*. Oxford: Oxford University Press, 2000.

Kornhuber, Hans Helmut, and Lüder Deecke. *The Will and Its Brain: An Appraisal of Reasoned Free Will*. Lanham, Md.: University Press of America, 2012.

Mischel, Walter. *The Marshmallow Test: Mastering Self-Control*. New York: Little, Brown, 2014.

Reinhart, Carmen, and Kenneth Rogoff. *This Time Is Different: Eight Centuries of Financial Folly.* Princeton, N.J.: Princeton University Press, 2009.

Schwartz, Jeffrey M., and Sharon Begley. *The Mind and the Brain: Neuroplasticity and the Power of Mental Force.* New York: Harper, 2002.

Schwartz, Jeffrey M., and Rebecca Gladding. *You Are Not Your Brain: The Four-Step Solution for Changing Bad Habits, Ending Unhealthy Thinking, and Taking Control of Your Life.* New York: Avery, 2011.

Senge, Peter, *The Fifth Discipline: The Art and Practice of the Learning Organization.* 2d ed. New York: Doubleday, 2006.

Senge, Peter, Art Kleiner, Charlotte Roberts, Richard Ross, George Roth, and Bryan Smith. *The Dance of Change: The Challenges to Sustaining Momentum in Learning Organizations.* New York: Doubleday, 1999.

Smith, Adam. *Theory of Moral Sentiments.* 6th ed. Ed. D. D. Raphael and A. L. Macfie. Indianapolis: Liberty Fund, 1982. [1st ed. pub. 1759; 6th ed. pub. 1790]

Index